CONTROL AND DYNAMIC SYSTEMS

Advances in Theory and Applications

Volume 26

ACADEMIC PRESS RAPID MANUSCRIPT REPRODUCTION

CONTROL AND DYNAMIC SYSTEMS

ADVANCES IN THEORY AND APPLICATIONS

Edited by
C. T. LEONDES

School of Engineering and Applied Science
University of California, Los Angeles
Los Angeles, California

VOLUME 26: SYSTEM IDENTIFICATION AND ADAPTIVE CONTROL

Part 2 of 3

ACADEMIC PRESS, INC.
Harcourt Brace Jovanovich, Publishers
San Diego New York Berkeley Boston
London Sydney Tokyo Toronto

ACADEMIC PRESS, INC.
1250 Sixth Avenue, San Diego, California 92101

United Kingdom Edition published by
ACADEMIC PRESS INC. (LONDON) LTD.
24–28 Oval Road, London NW1 7DX

LIBRARY OF CONGRESS CATALOG CARD NUMBER: 64-8027

ISBN 0–12–012726–1 (alk. paper)

PRINTED IN THE UNITED STATES OF AMERICA

87 88 89 90 9 8 7 6 5 4 3 2 1

CONTENTS

PREFACE

Volume 26 in this series is the second volume in the trilogy "Advances in the Theory and Application of System Parameter Identification and Adaptive Control." System parameter identification and adaptive control techniques have now matured to the point where such a trilogy is most timely for practitioners in the field who want a comprehensive reference source of techniques with significant applied implications.

The first contribution in this volume, "Techniques for Identification of Linear Time-Invariant Multivariable Systems," by C. Chen, presents a powerful, useful transformation technique for the identification of discrete time and continuous time linear time-invariant multivariable systems under rather general conditions. The noise-free case is emphasized, and the extension to noisy measurements is briefly discussed. The next contribution, "Techniques for the Selection of Identifiable Parameterizations for Multivariable Linear Systems," by M. Gevers and V. Wertz, is a comprehensive account of the use of identifiable representations for linear multivariable systems which includes a survey of methods for the estimation of these representations from measurable data. As a result, this contribution constitutes a useful reference source in the literature. "Parametric Methods for Identification of Transfer Functions of Linear Systems," by L. Ljung, the next contribution, treats the important subject of representing or developing a model for the true system by a representation which involves optimal choices of system parameters or design variables such that the output of the system representation is as close as possible to the output of the true system. In the next contribution, "Techniques in Dynamic Systems Parameter—Adaptive Control," R. Iserman points out the distinction between model reference adaptive control and model identification control, the two basic directions that have been followed in adaptive control over the years. Following this, a comprehensive review and analysis is given of model identification adaptive control. The resulting control systems are called parameter-adaptive control systems, the subject of this contribution. "Estimation of Transfer Function Models Using Mixed Recursive and Nonrecursive Methods," by M. Sherif and L. Liu, is next and focuses on three topics: (1) the technique for estimating the system and noise parameters, (2) the method for estimating the system and noise parameters, and (3) the start-up of the iterative procedure for system identification. The next contribution, "Techniques for Multivariable Self-Tuning Control," by H. T. Toivonen, presents the techniques for on-line system identification in combination with controller design methods and, as such, deals with one of the most powerful options for effective adaptive control. "A Covariance Control Theory," by A. Hotz and R. Skelton, presents covariance analysis as a key tool in the development of techniques in systems theory in the three areas of (1) system identification, (2) system state estimation, and (3) system model reduction. This unified treatment is rare in the literature and will prove to be a valuable reference source. The final contribution, "Adaptive Control with Recursive Identification for Stochastic Linear Systems," by H. Chen and L. Guo, presents the authors' powerful techniques for adaptive control by which both strong consistency of the estimate and optimality of adaptive tracking can be achieved simultaneously.

When the theme for this trilogy of volumes was decided upon, there seemed little doubt that it was most timely. The field has been quite active over nearly three decades and has now reached a level of maturity clearly calling for such a trilogy. Because of the substantially important contributions of the authors, however, all volumes promise to be not only timely but of substantial and lasting fundamental value.

CONTROL AND
DYNAMIC SYSTEMS

*Advances in Theory
and Applications*

Volume 26

Techniques for Identification of Linear Time-Invariant Multivariable Systems

CHI-TSONG CHEN

Department of Electrical Engineering
State University of New York
Stony Brook, New York 11794

I. INTRODUCTION

Identification of systems from measured input-output data is an important engineering problem and has been extensively studied [1-6]. A large number of techniques are available to identify systems. Some use recursive equations; some use non-recursive equations. Some are developed for noisy data; some for noise-free data. Some are for on-line computation and some are not.

In this article we discuss first a method to identify a linear time-invariant multivariable discrete-time system from noise-free input-output data. The system is assumed to be lumped and causal and is, therefore, describable by a proper rational matrix. The method does not require any prior knowledge of the structure of the system, such as its row degrees or observability indices. The method is applicable no matter whether the system is stable or not, is minimum phase or not, or is initially relaxed or not. The identification is achieved by transforming, using a numerically stable method, a matrix formed from the input-output data into a Hermite form. The result immediately yields a polynomial fraction description of the system. The polynomial fraction description is coprime and is in the echelon form, and thus an irreducible canonical-form state-variable description can also be obtained with essentially no additional computation.

In this article we also discuss persistent excitation of input sequences. A necessary and sufficient condition for the single-variable case is developed. We remark that even if none of the modes of a single-variable system is excited, an input sequence can still be persistently exciting and be used to identify the system. Some sufficient conditions for the multivariable case are discussed.

The result in the discrete-time case is then used to identify continuous-time systems. The extension of the method to noisy measurements is briefly discussed.

In this article the degree of a polynomial or a rational function is abbreviated as deg. If $f(z)$ is a polynomial, deg $f(z)$ is equal to its highest power. If $f(z)$ is a rational function, deg $f(z)$ denotes the degree of the denominator after canceling all common factors between the denominator and numerator.

II. PRELIMINARY

Consider a linear time-invariant discrete-time causal system with p input terminals and q output terminals. Its input and output sequences are denoted by $p \times 1$ vector $u(n) := u(nT)$ and $q \times 1$ vector $y(n) := y(nT)$, where T is the sampling period and the integer n ranges from $-\infty$ to ∞. It is assumed that the system is in continuous operation and the input and output data are measured from $n = n_1$ to n_2. Thus the problem is to identify the system from $\{u(n), y(n); n_1 \leq n \leq n_2\}$. If the system is known to be relaxed at n_1, we may assume, without loss of generality, that $u(n) = 0$ and $y(n) = 0$ for $n < n_1$.

The mathematical equation to be developed from the input-output data must be an input-output description or a transfer matrix. Of course, if the transfer matrix is identified, a state-variable description can then be obtained. Let the $q \times p$ proper rational matrix G(z) be the transfer matrix of the system. We express $\hat{G}(z)$ in the polynomial fractional form as

$$\hat{G}(z) = D^{-1}(z)N(z), \tag{1}$$

where D(z) and N(z) are, respectively, $q \times q$ and $q \times p$ polynomial matrices expressible as

$$\begin{aligned} D(z) &= D_0 + D_1 z + \cdots + D_k z^k, \\ N(z) &= N_0 + N_1 z + \cdots + N_k z^k, \end{aligned} \tag{2}$$

where D_i and N_i are $q \times q$ and $q \times p$ constant matrices. Because the system is assumed to be causal, the highest degree of N(z) is at most equal to that of D(z). If the input sequence $\{u(n), n = -\infty \text{ to } \infty\}$ is applied to the system, the output is equal to

$$\hat{y}(z) = \hat{G}(z)\hat{u}(z) = D^{-1}(z)N(z)\hat{u}(z), \tag{3}$$

where $\hat{u}(z)$ and $\hat{y}(z)$ are the two-sided z transforms of $\{u(n)\}$ and $\{y(n)\}$; that is,

$$\hat{u}(z) = \sum_{n=-\infty}^{\infty} u(n)z^{-n}, \quad \hat{y}(z) = \sum_{n=-\infty}^{\infty} y(n)z^{-n}. \tag{4}$$

We rewrite (3) as

$$D(z)\hat{y}(z) = N(z)\hat{u}(z),$$

or

$$[-N(z) \quad D(z)]\begin{bmatrix} u(z) \\ y(z) \end{bmatrix} = 0. \tag{5}$$

The substitution of (2) and (4) into (5) yields

$$[-N_0 \quad D_0 \quad -N_1 \quad D_1 \quad \cdots \quad -N_k \quad D_k]$$

$$\times \left[\begin{array}{cccccc} \cdots & u(n_1) & u(n_1+1) & \cdots & u(n_1+m) & \cdots \\ \cdots & y(n_1) & y(n_1+1) & \cdots & y(n_1+m) & \cdots \\ \hline \cdots & u(n_1+1) & u(n_1+2) & \cdots & u(n_1+m+1) & \cdots \\ \cdots & y(n_1+1) & y(n_1+2) & \cdots & y(n_1+m+1) & \cdots \\ \hline & \vdots & & & & \\ \hline \cdots & u(n_1+k) & u(n_1+k+1) & \cdots & u(n_1+m+k) & \cdots \\ \cdots & y(n_1+k) & y(n_1+k+1) & \cdots & y(n_1+m+k) & \cdots \end{array}\right]$$

$$=: MS_k(-\infty, \infty) = 0, \tag{6}$$

where $M := [-N_0 \; D_0 \; -N_1 \; D_1 \; \cdots \; -N_k \; D_k]$ is a $q \times (p + q)(k + 1)$ matrix and $S_k(-\infty, \infty)$ has $(p + q)(k + 1)$ rows and an infinite number of columns. The equation holds for any input sequence. In the identification problem from the input-output data, the matrix M is unknown and $S_k(-\infty, \infty)$ is known, and the problem becomes solving M from (6).

There are infinitely many equations in (6), far too many for solving M. In practice, we use only part of (6).

Define

$$S_k(n_1, n_2)$$

$$= \left[\begin{array}{cccc} u(n_1) & u(n_1 + 1) & \cdots & u(n_2) \\ y(n_1) & y(n_1 + 1) & \cdots & y(n_2) \\ \hline u(n_1 + 1) & u(n_1 + 2) & \cdots & u(n_2 + 1) \\ y(n_1 + 1) & y(n_1 + 2) & \cdots & y(n_2 + 1) \\ \hline \vdots & & & \\ \hline u(n_1 + k) & u(n_1 + k + 1) & \cdots & u(n_2 + k + 1) \\ y(n_1 + k) & y(n_1 + k + 1) & \cdots & y(n_2 + k + 1) \end{array}\right] \begin{array}{l} \text{one block} \\ \\ \\ \\ \\ \\ \\ \text{(k + 1) blocks} \end{array} . \tag{7}$$

It is a $(p + q)(k + 1) \times (n_2 - n_1 + 1)$ matrix. The input-output data used in the matrix range from $n = n_1$ to $n = n_2 + k + 1$. Now we solve M from

$$MS_k(n_1, n_2) = 0. \tag{8}$$

There are q rows in M. In order to find q nontrivial row solutions, we must find q linearly dependent rows in $S_k(n_1, n_2)$. Clearly, the highest degree, k, of D(z) and N(z) should be as small as possible; thus we search linearly dependent rows of S_k in order from top to bottom. In the next section, we discuss a method to carry out this search.

Before proceeding, we discuss the row degrees of a proper rational transfer matrix. Let $\hat{G}(s)$ be factored as $D^{-1}(z)N(z)$, where D(z) and N(z) are left coprime and D(z) is row reduced. Then the highest degree, ν_i, of the ith row of D(z) is called the ith row degree of $\hat{G}(z)$. There are q row degrees, where q is the number of output terminals. The set of row degrees $\{\nu_i, i = 1, 2, \ldots, q\}$ is an intrinsic property of $\hat{G}(z)$ and is independent of any particular coprime polynomial fraction

$D^{-1}(z)N(z)$ [7]. We call $\nu = \max\{\nu_i, i = 1, 2, \ldots, q\}$ the largest row degree of $\hat{G}(z)$. The degree of $\hat{G}(z)$ or the McMillan degree of $\hat{G}(z)$ is equal to $(\nu_1 + \nu_2 + \cdots + \nu_q)$. Thus the largest row degree of $\hat{G}(z)$ could be considerably smaller than the degree of $\hat{G}(z)$. Dual to the row degrees, using a right coprime fraction $N_r(z)D_r^{-1}(z)$ instead of a left coprime fraction $D^{-1}(z)N(z)$, a set of column degrees $\{\mu_i, i = 1, 2, \ldots, p\}$ can be defined. For the single-variable case, the row degree and the column degree reduce simply to the degree. The set of row degrees is essential in any identification scheme. We discuss in the following three more methods of computing the set.

1. Find an irreducible realization of $\hat{G}(z)$. Then the set of observability indices is equal to the set of row degrees of $\hat{G}(z)$ [7, pp. 199, 284].

2. Find a right coprime fraction $N_r(z)D_r^{-1}(z)$, not necessarily coprime, of $\hat{G}(z)$. Use the coefficient matrices of $N_r(z)$ and $D_r(z)$ to form a generalized resultant. By searching the linearly independent rows of the resultant in order from top to bottom, we will obtain a set of row indices. The set of row indices is equal to the set of row degrees [7, p. 616].

3. We expand $\hat{G}(z)$ as

$$\hat{G}(z) = H(0) + H(1)z^{-1} + H(2)z^{-2} + \cdots$$

where $H(i)$ are $q \times p$ constant matrices, and form the Hankel matrix

$$\overline{T} = \begin{bmatrix} H(1) & H(2) & H(3) & \cdots & H(k) \\ H(2) & H(3) & H(4) & \cdots & H(k+1) \\ \vdots & \vdots & \vdots & & \\ H(k) & H(k+1) & H(k+2) & \cdots & H(2k-1) \end{bmatrix}.$$

Note that $H(0)$ is not used in the Hankel matrix. Search of the linearly independent rows of $\overline{T}$ in order from top to bottom will yield the set of observability indices and consequently the set of row degrees [7, p. 261].

It is important to remark at this point that the formation of a Hankel matrix is not unique. For example, we may expand the ijth entry of $\hat{G}(z)$ as

$$\hat{g}_{ij}(z) = h_{ij}(0) + h_{ij}(1)z^{-1} + h_{ij}(2)z^{-2} + \cdots$$

and form a Hankel matrix H_{ij} for $\hat{g}_{ij}(z)$. We can then form the Hankel matrix

$$T = \begin{bmatrix} H_{11} & H_{12} & \cdots & H_{1p} \\ H_{21} & H_{22} & \cdots & H_{2p} \\ \vdots & \vdots & & \vdots \\ H_{q1} & H_{q2} & & H_{qp} \end{bmatrix}$$

for $\hat{G}(z)$. Now, if we search the linearly independent rows of T in order from top to bottom, we will not obtain the set of row degrees of $\hat{G}(z)$ [7, p. 272].

An implication of the preceding discussion is that the order of arrangement of u and y in S_k in (7) is crucial. If we grouped all u rows on the top or all y rows on the top, what will be discussed in the subsequent sections might not be applicable. Thus we use exclusively the form in (7). If the degree of a system is known a priori, it is permitted to group all y on the top, as is done in the literature.

III. ROW SEARCHING AND THE ECHELON-FORM SOLUTION

We discuss in this section the search of linearly dependent rows, in order from top to bottom, of matrix A. The row-searching algorithm discussed in [7] can be applied to carry out this search. The algorithm, however, is not necessarily numerically stable. Thus we discuss an alternative method in the following. It is achieved by first transforming A into a *staircase* form and then into a Hermite form by using only column operations [8]. Once A is transformed into a Hermite form, not only can the linearly dependent rows of A be obtained, but also the coefficients of the *unique* combination of every dependent row in terms of its previous independent rows of A can be read out.

The matrix A can be transformed into a staircase matrix by the following algorithm using exclusively elementary column operations.

Step 1. Let M = A. If M = 0, go to step 5.

Step 2. Find the first nonzero row of M from the top. Find the largest element in magnitude in that row and move it to the first column. Normalize the element to 1 and call it the pivot.

Step 3. Use the pivot to eliminate all elements on its right-hand side.

Step 4. Delete the first column, the row with the pivot, and the zero rows above it. Rename the remaining matrix as M and go to step 1.

Step 5. Stop.

If A is a 7×8 matrix, at the end of the algorithm A will be transformed into the form

$$\bar{A} = \begin{bmatrix} 0 & 0 & 0 & 0 & 0 & 0 & 0 & 0 \\ \textcircled{1} & 0 & 0 & 0 & 0 & 0 & 0 & 0 \\ x & \textcircled{1} & 0 & 0 & 0 & 0 & 0 & 0 \\ x & x & 0 & 0 & 0 & 0 & 0 & 0 \\ x & x & \textcircled{1} & 0 & 0 & 0 & 0 & 0 \\ x & x & x & 0 & 0 & 0 & 0 & 0 \\ x & x & x & 0 & 0 & 0 & 0 & 0 \end{bmatrix}, \qquad (9)$$

where x denotes possible nonzero elements. We call such a matrix a staircase matrix. It is clear that the rows with pivot 1 are linearly independent of their previous rows. The rows without pivot are linearly dependent on their previous rows. Thus if a matrix is transformed into the staircase matrix by elementary column operations, the linearly dependent rows, in order from top to bottom, of the matrix can be readily obtained.

With additional elementary column operations, the matrix in (9) can be transformed into the form

$$\bar{\bar{A}} = \begin{bmatrix} 0 & 0 & 0 & 0 & 0 & 0 & 0 & 0 \\ \textcircled{1} & 0 & 0 & 0 & 0 & 0 & 0 & 0 \\ 0 & \textcircled{1} & 0 & 0 & 0 & 0 & 0 & 0 \\ a_{41} & a_{42} & 0 & 0 & 0 & 0 & 0 & 0 \\ 0 & 0 & \textcircled{1} & 0 & 0 & 0 & 0 & 0 \\ a_{61} & a_{62} & a_{63} & 0 & 0 & 0 & 0 & 0 \\ a_{71} & a_{72} & a_{73} & 0 & 0 & 0 & 0 & 0 \end{bmatrix}. \tag{10}$$

It is called a column echelon form or a column Hermite form. It is achieved by using the last pivot to eliminate, using elementary column operations, its left-hand-side elements and then using the pivot next to the last to eliminate *its* left-hand-side elements. Proceeding upward, the Hermite form can be obtained.

Let A_i be the ith row of A. From (10) we see that A_1, A_4, A_6, and A_7 are linearly dependent on their previous rows and can be written as linear combinations of their previous rows. In general, the combinations are not unique. For example, we have

$$A_4 = a_{41}A_2 + a_{42}A_3 = \alpha A_1 + a_{41}A_2 + a_{42}A_3$$

for any α and

$$A_6 = a_{61}A_2 + a_{62}A_3 + a_{63}A_5$$
$$= \alpha\beta A_1 + (a_{61} + a_{41}\beta)A_2 + (a_{62} + a_{42}\beta)A_3 - \beta A_4 + a_{63}A_5$$

for any α and β. However, if we require the coefficients of combination associated with dependent rows A_1 and A_4 to be zero, or equivalently, we express a dependent row exclusively in terms of its previous linearly independent rows, then the combination is unique. Furthermore, the coefficients of the unique combination can be read out directly from (10). For example, we have

$$A_4 = a_{41}A_2 + a_{42}A_3,$$

$$A_6 = a_{61}A_2 + a_{62}A_3 + a_{63}A_5,$$

and

$$A_7 = a_{71}A_2 + a_{82}A_3 + a_{73}A_5.$$

These three equations together with $A_1 = 0$ can be written as

$$NA := \begin{bmatrix} \textcircled{1} & 0 & 0 & 0 & 0 & 0 & 0 \\ 0 & -a_{41} & -a_{42} & \textcircled{1} & 0 & 0 & 0 \\ 0 & -a_{61} & -a_{62} & 0 & -a_{63} & \textcircled{1} & 0 \\ 0 & -a_{71} & -a_{72} & 0 & -a_{73} & 0 & \textcircled{1} \end{bmatrix} \begin{bmatrix} A_1 \\ A_2 \\ A_3 \\ A_4 \\ A_5 \\ A_6 \\ A_7 \end{bmatrix} = \begin{bmatrix} 0 \\ 0 \\ 0 \\ 0 \end{bmatrix}. \qquad (11)$$

We see that the structure of N is similar to the one in (10) and is said to be in a row echelon form or a row Hermite form. The four pivots indicated by circles correspond to the four dependent rows of A. In fact, N is a basis of the left null space of A. The N in (11) will be called the echelon-form solution of NA = 0. The polynomial fraction to be identified in this article will be in the echelon form.

Steps 2 and 3 of the algorithm that transform a matrix into a staircase form are actually Gaussian elimination with partial pivoting. It is, in practice, numerically stable. The Householder transformation can also be used to carry out these steps. In this case, the pivot is equal to the norm of each row. The pivot can then be normalized in the subsequent step. The transformation of the staircase matrix into the Hermite form essentially carries out back substitutions. This is a widely used algorithm and is a numerically acceptable method [9, p. 150]. Thus the computation discussed in this section is, in practice, numerically stable and can be easily programmed.

The search of linearly dependent rows is basically the same as the computation of the rank of a matrix and is an ill-conditioned numerical problem. The singular value decomposition is considered to be the most reliable method of computing the rank. However, it is expensive and comparable results can often be obtained by using the Householder transformation with pivoting [19,7]. The singular value decomposition and the Householder transformation with pivoting may require the change of the ordering of rows and cannot be used in our row searching. Thus the Householder transformation (without pivoting) may be the best method available at present to carry out the row searching. However, in our limited experience, the differences between the results obtained by using the Gaussian elimination with partial pivoting and the Householder transformation are negligible.

The Hermite (normal) form or echelon form is extensively discussed in [8]. The relationship between (10) and (11), however, is not explicitly developed in [8] and, to the best of my knowledge, is not discussed elsewhere. Thus it is developed in detail in this section.

IV. PERSISTENT EXCITATION OF INPUT SEQUENCES

It is well known that the input sequence to a system must be persistently exciting in order to identify the system. This fundamental result was first established in [10] and has since appeared in a great number of identification articles. Significant results are reported in, among others, references [11] and [12].

In this section we introduce persistent excitation in a slightly different way. Some conditions for persistent excitation will also be developed. The development relies heavily on the realization on Hankel matrices [7].

Consider $S_k(n_1, n_2)$ defined in (7). We require $n_2 - n_1 + 1 \geq (p + q)(k + 1)$ to ensure that $S_k(n_1, n_2)$ has more columns than rows. We call the rows in $S_k(n_1, n_2)$ formed from $\{u(n)\}$ u rows and those from $\{y(n)\}$ y rows. Thus each block of $S_k(n_1, n_2)$ has p number of u rows and q number of y rows. It is assumed that the linearly dependent rows of S_k, in order from top to bottom, have been found. Then the input sequence $\{u(n)\}$ is said to be persistently exciting if all u rows in $S_k(n_1, n_2)$ are linearly independent (of their previous rows). Clearly, the persistent excitation of an input sequence depends on the three integers k, n_1, and n_2. If an upper bound of the degree α of the system is available, we may set n_2 equal to $n_1 + (p + q)(\alpha + 1)$. If no such bound is available, theoretically, we should set $n_2 = \infty$. This is not possible and a finite n_2 is always used in practice. In no case should n_2 be smaller than $n_1 - 1 + (p + q)(k + 1)$. For completeness, we set $n_2 = \infty$ in the following discussion.

We discuss now the determination of k required in the identification. First, we search the linearly dependent rows of $S_0(n_1, \infty)$ in order from top to bottom. If linearly dependent rows appear in u rows, the input sequence is not persistently exciting and a different input sequence must be used. Under the assumption that all u rows are linearly independent, if not all q number of y rows in S_0 are linearly dependent, we increase k by one and search $S_1(n_1, \infty)$. If not all q y rows of the *last* block of $S_1(n_1, \infty)$ are linearly dependent, we increase k by one and search $S_2(n_1, \infty)$. We stop the search if all q number of y rows in the last block of $S_k(n_1, \infty)$ are linearly dependent on their previous rows. This k can be shown to be equal to ν, the highest row degree of G(z) [7]. In other words, ν is the least integer such that all q number of y rows in the last block of $S_\nu(n_1, \infty)$ are linearly dependent. Once ν is found, it is not necessary to carry out any further searching.

We develop in the following the condition for $\{u(n)\}$ to be persistently exciting in $S_\nu(n_1, \infty)$. Because of time invariance, we may assume, without loss of generality, that $n_1 = 1$ and consider $S_\nu(1, \infty)$ and $\{u(n), n = 1, 2, 3, \ldots\}$. Let $\hat{u}(z)$ be the z transform of $\{u(n), n = 1, 2, 3, \ldots\}$. Then the matrix

$$S_\nu(1, \infty) = \left[\begin{array}{cccc} u(1) & u(2) & u(3) & \cdots \\ y(1) & y(2) & y(3) & \cdots \\ \hline u(2) & u(3) & u(4) & \cdots \\ y(2) & y(3) & y(4) & \cdots \\ \hline \vdots & \vdots & \vdots & \\ \hline u(\nu + 1) & u(\nu + 2) & u(\nu + 3) & \cdots \\ y(\nu + 1) & y(\nu + 2) & y(\nu + 3) & \cdots \end{array}\right]$$

is the Hankel matrix of

$$\begin{bmatrix} \hat{u}(z) \\ \hat{y}(s) \end{bmatrix}. \tag{12}$$

We discuss now the reason of choosing $n_1 = 1$ rather than $n_1 = 0$. Let $\hat{u}(z)$ be the z transform of $\{u(n), n = 0, 1, 2, \ldots\}$. Then $S_\nu(0, \infty)$ is *not* the Hankel matrix of $[\hat{u}'(z)\hat{y}'(z)]'$, where the prime denotes the transpose, because $u(0)$ and $y(0)$ are not included in the Hankel matrix. Thus it is more convenient to choose $n_1 = 1$ than $n_1 = 0$.

In order for $\{u(n), n = 1, 2, \ldots\}$ to be persistently exciting, it is necessary that the Hankel matrix of $\hat{u}(z)$ in $S_\nu(1, \infty)$ have a full row rank by itself. A necessary and sufficient condition for the Hankel matrix to have a full rank is that all row degrees of $\hat{u}(z)$ be equal to or larger than $\nu + 1$. To find the row degrees of $\hat{u}(z)$, we must compute a left coprime fraction. This is complicated. Simpler sufficient conditions for the Hankel matrix of $\hat{u}(z)$ in $S_\nu(1, \infty)$ to have a full rank are that, for $i = 1, 2, \ldots, p$,

$$\deg \hat{u}_i(z) \geq \nu + 1$$

and that the set of the poles of $\hat{u}_i(z)$ are mutually disjoint, where $\hat{u}_i$ is the ith component of $\hat{u}(z)$. These conditions, in turn, are necessary conditions for $\{u(n), n = 1, 2, 3, \ldots\}$ to be persistently exciting in $S_\nu(1, \infty)$.

For the single-input single-output case, we have the following theorem.

Theorem 1

Consider a single-variable system with transfer function $\hat{g}(z) = N(z)/D(z)$, where $N(z)$ and $D(z)$ are coprime and

$\deg N(z) \leq \deg D(z) = \nu$. If the z transform $\hat{u}(z)$ of $\{u(n),$ $n = 1, 2, 3, \ldots\}$ is a rational function, then $\{u(n)\}$ is persistently exciting in $S_\nu(1, \infty)$ if and only if

$$\deg\left(\frac{1}{D(z)}\,\hat{u}(z)\right) \geq 2\nu + 1.$$

Before proving the theorem, we discuss first its implication. Let $\hat{u}(z) = N_u(z)/D_u(z)$, where $N_u(z)$ and $D_u(z)$ are coprime. The degree of the rational function

$$\frac{1}{D(z)}\,\hat{u}(z) = \frac{N_u(z)}{D(z)D_u(z)}$$

is equal to the degree of its denominator after canceling the common factor between $D(z)$ and $N_u(z)$. Let $\hat{D}(z)$ be the greatest common factor or divisor (GCD) of $D(z)$ and $N_u(z)$; that is, $D(z) = \bar{D}(z)\hat{D}(z)$ and $N_u(z) = \bar{N}_u(z)\hat{D}(z)$. Let $\deg \hat{D}(z) = k$. Then we have

$$\deg\left(\frac{1}{D(z)}\,\hat{u}(z)\right) = \deg\left[\frac{N_u(z)}{D(z)D_u(z)}\right] = \left[\frac{\bar{N}_u(z)}{\bar{D}(z)D_u(z)}\right]$$

$$= \deg \bar{D}(z) + \deg D_u(z) = \nu - k + \deg D_u(z).$$

Thus $\{u(n), n = 1, 2, 3, \ldots\}$ is persistently exciting in $S_\nu(1, \infty)$ if and only if

$$\deg D_u(z) \geq 2\nu + 1 - \nu + k = \nu + k + 1,$$

where k is the degree of the GCD of $D(z)$ and $N_u(z)$ and is at most equal to ν. Thus if $k = 0$, that is, if there is no cancellation in $N_u(z)/D(z)$, then $\{u(n), n = 1, 2, 3, \ldots\}$ is persistently exciting in $S_\nu(1, \infty)$ if and only if $\deg \hat{u}(z) \geq \nu + 1$. If $\deg \hat{u}(z) \geq 2\nu + 1$, then $\{u(n), n = 1, 2, 3, \ldots\}$ is always persistently exciting in $S_\nu(1, \infty)$ no matter whether there are cancellations in $N_u(z)/D(z)$ or not. If the z transform of $\{u(n), n \geq 1\}$ is not a rational function, $\hat{u}(z)$ may be considered to have a degree of infinity and is therefore persistently exciting. If an input sequence is generated randomly, the probability of its z transform being a rational function is almost zero. Thus the probability of a randomly generated input sequence being persistently exciting is almost one.

The condition of persistent excitation also has the following interesting interpretation. If there is no cancellation between $N_u(z)$ and $D(z)$, then the tandem connection of $\hat{u}(z)$ followed by $\hat{g}(z)$ is controllable and all the modes of the system are excited by the input. In this case, the input sequence is persistently exciting if and only if $\deg \hat{u}(z) \geq \nu + 1$. To identify a system, it is, however, not necessary to excite all the modes of the system. For example, if $D(z) = N_u(z)$, then none of the modes of the system is excited at the output $\{y(n)\}$. However, in this case, if $\deg \hat{u}(z) \geq 2\nu + 1$, the input sequence is persistently exciting and the system can be identified from the input-output data.

Now we are ready to establish Theorem 1. The idea is to show that the first row degree of $[\hat{u}(z)\ \hat{y}(z)]'$ is at least $(\nu + 1)$ under the condition of the theorem and thus that all u rows in $S_\nu(1, \infty)$ are linearly independent of their previous rows. It is important to note that, although $\deg \hat{u}(z) \geq \nu + 1$, there is no guarantee that the first row degree of $[\hat{u}(z)\ \hat{y}(z)]'$ is at least $\nu + 1$. For example, if $\hat{g}(z) = 1/(z + 1)$, $\hat{u}(z) = (z + 1)/z^2$, then $\hat{y}(z) = 1/z^2$. Using the method in [7], we can readily compute the coprime fraction

$$\begin{bmatrix} \hat{u}(z) \\ \hat{y}(z) \end{bmatrix} = \begin{bmatrix} \dfrac{z+1}{z^2} \\ \dfrac{1}{z^2} \end{bmatrix} = \begin{bmatrix} z-1 & 1 \\ -1 & z+1 \end{bmatrix}^{-1} \begin{bmatrix} 1 \\ 0 \end{bmatrix}.$$

Although $\deg \hat{u}(z) = 2$, the first row degree of $[\hat{u}(z)\ \hat{y}(z)]'$ is 1.

Proof of Theorem 1. Consider

$$F(z) := \begin{bmatrix} \hat{u}(z) \\ \hat{y}(z) \end{bmatrix} = \begin{bmatrix} \dfrac{N_u(z)}{D_u(z)} \\ \dfrac{N(z)}{D(z)} \cdot \dfrac{N_u(z)}{D_u(z)} \end{bmatrix} = \begin{bmatrix} \dfrac{N_u(z)}{D_u(z)} \\ \dfrac{\bar{N}(z)\hat{N}(z)}{\bar{D}(z)\hat{D}(z)} \cdot \dfrac{\bar{N}_u(z)\hat{D}(z)}{\bar{D}_u(z)\hat{N}(z)} \end{bmatrix}$$

$$= \begin{bmatrix} \dfrac{N_u(z)}{D_u(z)} \\ \dfrac{\bar{N}(z)}{\bar{D}(z)} \cdot \dfrac{\bar{N}_u(z)}{\bar{D}_u(z)} \end{bmatrix},$$

where $\hat{N}(z)$ is the GCD of $N(z)$ and $D_u(z)$. The least common denominator of $\hat{u}(z)$ and $\hat{y}(z)$ is $\bar{D}(z)D_u(z)$. It is the characteristic polynomial of $F(z)$ [7, p. 235]. We write

$$F(z) = \begin{bmatrix} N_u(z)\bar{D}(z) \\ \bar{N}(z)\bar{N}_u(z)\hat{N}(z) \end{bmatrix} \left[\bar{D}(z)D_u(z)\right]^{-1}$$

$$= \begin{bmatrix} \bar{N}_u(z)D(z) \\ \bar{N}_u(z)N(z) \end{bmatrix} \left[\bar{D}(z)D_u(z)\right]^{-1} = \begin{bmatrix} X(z) & Y(z) \\ -N(z) & D(z) \end{bmatrix}^{-1} \begin{bmatrix} \bar{N}_u(z) \\ 0 \end{bmatrix}$$

$$=: A^{-1}(z)B(z) \tag{13}$$

where the polynomials $X(z)$ and $Y(z)$ are to be solved from the Diophantine equation

$$X(z)D(z) + Y(z)N(z) = \bar{D}(z)D_u(z). \tag{14}$$

Now we prove the sufficiency of the theorem. Let $\deg\left[\bar{D}(z)D_u(z)\right] = 2\nu + 1$. Because $D(z)$ and $N(z)$ are coprime and $\deg N(z) \le \deg D(z) = \nu$, polynomial solutions $X(z)$ and $Y(z)$ with $\deg Y(z) \le \deg X(z) = \nu + 1$ exist in (14) [7, p. 461]. Consider the polynomial matrices $A(z)$ and $B(z)$ defined in (13). Because of

$$\det A(z) = \deg[X(z)D(z) + Y(z)N(z)] = \deg\left[\bar{D}(z)D_u(z)\right]$$
$$= \text{the characteristic polynomial of } F(z),$$

the fraction $A^{-1}(z)B(z)$ is left coprime. Because the sum of the row degrees of $A(z)$ is equal to $\deg\left[\bar{D}(z)D_u(z)\right]$, $A(z)$ is row reduced. Thus the two row degrees r_1 and r_2 of $F(z)$ are

$$r_1 = \nu + 1 \quad \text{and} \quad r_2 = \nu.$$

This implies that all u rows in $S_\nu(1, \infty)$ are linearly independent [7]. This establishes the sufficiency of the theorem for $\deg\left[\bar{D}(z)D_u(z)\right] = 2\nu + 1$. The case $\deg\left[\bar{D}(z)D_u(z)\right] > 2\nu + 1$ can be similarly proved. If $\deg\left[\bar{D}(z)D_u(z)\right] \le 2\nu$, then the first row degree of $[u(z)\ y(z)]'$ is at most ν. Thus not all u rows in $S_\nu(1, \infty)$ can be linearly independent, and $\{u(n),\ n \ge 1\}$ is not persistently exciting. Q.E.D.

In employing $S_\nu(1, \infty)$ in the identification, the input signal must have a degree of at least one higher than that of the system. In the following we show that *any* input sequence is persistently exciting if a different data set is used. However, the system must be known to be initially relaxed. Recall that if a system is relaxed at $n_1 = 1$, we may assume $u(n) = 0$ and $y(n) = 0$ for $n < 1$.

Corollary 1

Consider a single-variable system of degree ν. If it is initially relaxed at $n_1 = 1$, then any nonzero input sequence $\{u(n),\ n = 1, 2, 3, \ldots\}$ is persistently exciting in $S_\nu(-d, \infty)$, with $d \ge \nu - 1$.

Proof. If the system is relaxed at $n_1 = 1$, then $u(n) = 0$ and $y(n) = 0$ for $n \le 0$. Under this assumption, it is permitted to use $S_\nu(-d, \infty)$, with $d \ge 0$, in the identification. If $\hat{u}(z)$

is the z transform of $\{u(n),\ n = 1, 2, 3, \ldots\}$, the z transform of

$$\{u_1(n)\} = \{\underbrace{0, 0, \ldots, 0}_{(d+1)\text{ zeros}}, u(1), u(2), u(3), \ldots\}$$

is $\hat{u}(z)/z^{d+1}$. It follows from Theorem 1 that $\{u_1(n)\}$ is persistently exciting in $S_\nu(-d, \infty)$ if and only if

$$\deg\left(\frac{1}{D(z)}\ \frac{\hat{u}(z)}{z^{d+1}}\right) = \deg\left(\frac{1}{D(z)} \cdot \frac{N_u(z)}{D_u(z)\,z^{d+1}}\right) \geq 2\nu + 1. \tag{15}$$

If the z transform of $\{u(n),\ n = 1, 2, \ldots\}$ is a rational function, we have $\deg D_u(z) \geq \deg N_u(z) + 1$. This implies

$$\begin{aligned}\deg\left(\frac{1}{D(z)} \cdot \frac{N_u(z)}{D_u(z)\,z^{d+1}}\right) &\geq \deg D(z) + \deg D_u(z)\\ &\quad - \deg N_u(z) + d + 1\\ &\geq \nu + d + 2.\end{aligned}$$

Thus if $d \geq \nu - 1$, for any nonzero $\{u(n),\ n = 1, 2, 3, \ldots\}$, the condition in (15) is satisfied. This proves the corollary. Q.E.D.

Example. Consider a system with transfer function $1/(z+1)$. It is assumed to be relaxed at $n_1 = 1$. Let $u(z) = z^{-1} + z^{-2}$. Then we have $\hat{y}(z) = \hat{g}(z)\hat{u}(z) = z^{-2}$ and

$$S_1(0, \infty) = \left[\begin{array}{cccc} u(0) & u(1) & u(2) & \cdots \\ y(0) & y(1) & y(2) & \cdots \\ \hline u(1) & u(2) & u(3) & \cdots \\ y(1) & y(2) & y(3) & \cdots \end{array}\right] = \left[\begin{array}{ccccc} 0 & 1 & 1 & 0 & \cdots \\ 0 & 0 & 1 & 0 & \cdots \\ \hline 1 & 1 & 0 & 0 & \cdots \\ 0 & 1 & 0 & 0 & \cdots \end{array}\right].$$

All u rows are linearly independent. Thus the input sequence is persistently exciting in $S_1(0, \infty)$. It is, however, not persistently exciting in $S_1(1, \infty)$.

For the single-variable case, Theorem 1 yields a necessary and sufficient condition for an input sequence to be persistently exciting. For the multivariable case, no such theorem is available at present. Although the following theorem is believed to be true in general, we are able to establish it only for cyclic transfer matrices [7, p. 469].

Theorem 2

Consider a multivariable system with $q \times p$ cyclic proper transfer matrix $\hat{G}(q) = D^{-1}(z)N(z) = N_r(z)D_r^{-1}(z)$, where D and N are left coprime and D is row reduced; D_r and N_r are right coprime and D_r is column reduced. Let $\hat{u}(z)$ be the z transform of $\{u(n), n = 1, 2, 3, \ldots\}$ and be a strictly proper rational vector. Let $\hat{u}_i(z)$, $i = 1, 2, \ldots, p$, be the ith component of $\hat{u}(z)$. If

(i) $\deg \hat{u}_i(z) \geq \nu + 1$, $i = 1, 2, \ldots, p$,

(ii) the sets of the poles of $\hat{u}_i(z)$, $i = 1, 2, \ldots, p$, are mutually disjoint, and

(iii) the tandem connection of $\hat{u}(z)$ followed by $\hat{G}(z)$ is controllable, then $\{u(n), n \geq 1\}$ is persistently exciting in $S_\nu(1, \infty)$.

Proof. Consider

$$F(z) := \begin{bmatrix} \hat{u}(z) \\ \hat{y}(z) \end{bmatrix} = \begin{bmatrix} D_u^{-1}(z)N_u(z) \\ D^{-1}(z)N(z)D_u^{-1}(z)N_u(z) \end{bmatrix}$$

$$= \begin{bmatrix} D_u(z) & 0 \\ -N(z) & D(z) \end{bmatrix}^{-1} \begin{bmatrix} N_u(z) \\ 0 \end{bmatrix} =: A^{-1}(z)B(z), \tag{16}$$

where $u(z) = D_u^{-1}(z)N_u(z)$ is coprime and $D_u(z)$ is row reduced. Under the assumptions of (i) and (ii), $D_u(z)$ has row degrees at least $\nu + 1$. We write

$$D^{-1}(z)N(z)D_u^{-1}(z)N_u(z) = N_r(z)D_r^{-1}(z)D_u^{-1}(z)N_u(z)$$
$$= N_r(z)[D_u(z)D_r(z)]^{-1}N_u(z).$$

If $\hat{u}(z)$ followed by $\hat{G}(z)$ is controllable, then $D_u(z)D_r(z)$ and $N_u(z)$ are left coprime. This implies that, roughly speaking, no poles of $\hat{G}(z)$ [or no roots of det $D_r(z)$ or det $D(z)$] are canceled in $\hat{G}(z)\hat{u}(z)$. Under this assumption and the cyclicity of $\hat{G}(z)$, the characteristic polynomial of $F(z)$ in (16) is det $D_u(z)$ · det $D(z)$. Consequently, the left fraction $A^{-1}(z)B(z)$ in (16) is coprime. The matrix $A(z)$ is row reduced as is implied by the row reducedness of $D_u(z)$ and $D(z)$. Thus the first p row degrees of $F(z)$ are equal to the row degrees of $\hat{u}(z)$. Consequently, all u rows are linearly independent in $S_\nu(1, \infty)$ and are persistently exciting. Q.E.D.

Similar to the scalar case, it is conjectured that if $\{u(n), n = 1, 2, \ldots\}$ is generated randomly, the input sequence will be persistently exciting. In any case, the persistent excitation of an input sequence is automatically checked in the process of identification.

We remark that the full row rank of the matrix

$$\begin{bmatrix} u(n_1) & u(n_1 + 1) & \cdots & u(n_2) \\ u(n_1 + 1) & u(n_1 + 2) & \cdots & u(n_2 + 1) \\ \vdots & \vdots & & \vdots \\ u(n_1 + \nu) & u(n_1 + \nu + 1) & \cdots & u(n_2 + \nu) \end{bmatrix}$$

is equivalent to the nonsingularity of

$$\sum_{i=n_1}^{n_2} \bar{u}(i)\bar{u}'(i),$$

where $\bar{u}'(i) := [u'(i)\ u'(i+1) \cdots u'(i+\nu)]$ and the prime denotes the transpose. The latter form is used in most papers in the discussion of persistent excitation.

Before proceeding, we remark that Theorem 1 and its corollary are developed for systems without any prior knowledge of degrees. If the degrees of the denominator and numerator are known, the condition of persistent excitation can be relaxed [18]. We also mention a degenerate case of Theorem 1. For example, if $\hat{g}(z) = z^2/(z+1)^2$ and $\hat{u}(z) = (z+1)^2/z^3$, then $\hat{y}(z) = 1/z$. According to Theorem 1, the input is not persistently exciting in $S_2(1, \infty)$. Indeed, if we form $S_2(1, \infty)$ as

$$\left[\begin{array}{ccccc} 1 & 2 & 1 & 0 & 0 \quad \cdots \\ 1 & 0 & 0 & 0 & 0 \\ \hline 2 & 1 & 0 & 0 & 0 \\ 0 & 0 & 0 & 0 & 0 \\ \hline 1 & 0 & 0 & 0 & 0 \\ 0 & 0 & 0 & 0 & 0 \end{array}\right],$$

the third u row is dependent and the input is not persistently exciting. However, if we search the linearly dependent rows of $S_k(1, \infty)$ in order from top to bottom, we should stop at $S_1(1, \infty)$, because the y row in its last block becomes dependent. Although $\{u(n)\}$ is persistently exciting in $S_1(1, \infty)$, the computed $\hat{g}(z)$ is zero. Thus this type of degenerate case should be excluded from Theorem 1.

V. IDENTIFICATION OF DISCRETE-TIME SYSTEMS

The identification of a system can be accomplished by solving $MS_k(n_1, n_2) = 0$. It will be achieved by searching the linearly dependent rows of $S_k(n_1, n_2)$ in order from top to bottom. We stop the search once all the q number of y rows in the last block of $S_k(n_1, n_2)$ become linearly dependent. The k yields the highest row degree, ν, of the system. Corresponding to the last q linearly dependent y rows in $S_\nu(n_1, n_2)$, q rows of nonzero solutions $M = [-N_0 \; D_0 \; -N_1 \; D_1 \cdots -N_\nu \; D_\nu]$ can be obtained. From M, we can compute D(z) and N(z) as in (2). Then $D^{-1}(z)N(z)$ is a mathematical description of the system. This description, however, is not necessarily coprime. In the following we introduce the concept of primary dependent rows from which a coprime description can be readily obtained.

There are $(\nu + 1)$ blocks in $S_\nu(n_1, n_2)$. In each block there are p u rows and q y rows. Let $\bar{y}_{ik}$ denote the ith y row in the kth block of $S_\nu(n_1, n_2)$ with $i = 1, 2, \ldots, q$ and $k = 1, 2, \ldots, \nu + 1$. For each i, the first $\bar{y}_i$ or, equivalently, the $\bar{y}_{ik}$ with the smallest integer k, call it $\nu_i + 1$, that becomes linearly dependent on its previous rows is called the *primary* dependent row. Because of the structure of $S_\nu(n_1, n_2)$, if $\bar{y}_{i(\nu_i+1)}$ becomes dependent, all $\bar{y}_{ik}$, $k = \nu_i + 2, \nu_i + 3, \ldots, \nu + 1$ will be dependent on their previous rows and are called nonprimary dependent rows. Thus ν_i is equal to the number of linearly independent $\bar{y}_i$ rows in $S_\nu(n_1, n_2)$. For each $i = 1, 2, \ldots, q$, there is a primary dependent row. Thus there are totally q primary dependent rows. Corresponding to these q primary dependent rows, we can find q nonzero rows of M. An M can be readily read out if $S_\nu(n_1, n_2)$ is transformed into the Hermite form as discussed in Section III. This M is in the row echelon form. The corresponding D(z) and N(z) have the following properties: (i) N(z) and D(z) are left coprime; (ii) D(z) is row reduced. Its row degrees yield the row degrees of the system, and

(iii) D(z) and N(z) are in the echelon form or Popov form. A proof of these assertions can be found in [7].

We give an example to illustrate the identification procedure. Consider the two-input/two-output system

$$x(n+1) = \begin{bmatrix} 0.5 & 1 & 0 \\ 0 & 0.5 & 0 \\ 0 & 0 & 1 \end{bmatrix} x(n) + \begin{bmatrix} 0 & 2 \\ 1 & 0 \\ 0 & 0.2 \end{bmatrix} u(n),$$
$$y(n) = \begin{bmatrix} 0.5 & -1 & 1 \\ 0.25 & 0 & 1 \end{bmatrix} x(n). \tag{17}$$

It is not a BIBO stable system because the system has one eigenvalue on the unit circle. The following computation is carried out on a Macintosh personal computer.

Let x(1) = [1 -1 2]'. A pseudorandom input sequence with amplitude ranging between ±5 is generated and the output is computed. We give a partial listing in the following.

	$n = 1$	2	...	14	15	16	...
$u_1(n)$	0.421	1.36		0.668	-3.696	-1.16	
$u_2(n)$	-1.11	1.40		1.932	-0.839	3.15	
$y_1(n)$	3.5	0.49		-1.786	2.711	6.06	
$y_2(n)$	2.25	1.09		1.060	3.376	2.72	

The exact and complete data can be found in Table I.

Now the problem is to dientify the system, without any prior knowledge of its row degrees, from the data {u(n), y(n), n = 1, 2, 3, ...}. We form $S_2(1, 14)$. For easy printout, we print the *transpose* of $S_2(1, 14)$ as the first matrix in Table I. We use the algorithm in Section III to transform $S_2(1, 14)$, by using column operations, into a staircase form and then into the

Hermite form. The transposes of the resulting matrices are printed in Table I. The columns with encircled pivot 1's are linearly independent of their left-hand-side columns. Or the corresponding rows of $S_2(1, 14)$ are linearly independent of their previous rows. In the computation, an element smaller than 10^{-10} is considered as zero. We see that the y_1 in the second block of $S_2(1, 14)$ is a primary dependent row, and so is the y_2 in the third block of $S_2(1, 14)$. The y_1 in the third block is a nonprimary dependent row. The two echelon-form solutions corresponding to the two primary dependent rows can be read from Table I as

$$\begin{bmatrix} 1 & -1.2 & \vdots & 0 & -1 & | & 0 & 0 & \vdots & \textcircled{1} & 0 & | & 0 & 0 & \vdots & 0 & 0 \\ -0.25 & 0.95 & \vdots & -0.25 & 1.25 & | & 0 & -0.7 & \vdots & \underline{0} & -2 & | & 0 & 0 & \vdots & \underline{0} & \textcircled{1} \end{bmatrix}$$

$$\times S_2(1, 14) = 0. \tag{18}$$

We explain how the second solution is obtained from the Hermite form of $S_2(1, 14)$. The element 1 indicated by the circle corresponds to the primary dependent y_2 row. The nine elements in the box in the last column of the Hermite matrix in Table I yield the coefficients, except the reversal of the signs, of the unique combination of the primary dependent y_2 row in terms of its previous linearly *independent* rows. The two underlined zeros are the coefficients corresponding to its previous two linearly *dependent* y_1 rows and do not appear in Table I. Thus the solutions in (18) can indeed be directly obtained from the Hermite form of $S_2(1, 14)$.

TABLE I.

1	u_2	y_1	y_2	u_1	u_2	y_1 [a]	y_2	u_1	u_2	y_1 [b]	y_2 [a]
.421142578	-1.11450191	3.5	2.25	1.36047363	1.40014648	.491455078	1.09484863	-.89111328	2.16918945	1.41455078	2.39636230
1.36047363	1.40014648	.491455078	1.09484863	-.89111328	2.16918945	1.41455078	2.39636230	4.35363769	.048217773	5.89050293	4.07543945
-.89111328	2.16918945	1.41455078	2.39636230	4.35363769	.048217773	5.89050293	4.07543945	-.54199218	-4.4342041	-.22033691	3.25930786
4.35363769	.048217773	5.89050293	4.07543945	-.54199218	-4.4342041	-.22033691	3.25930786	-2.0721435	2.25036621	-1.5197448	.835601806
-.54199218	-4.4342041	-.22033691	3.25930786	-2.0721435	2.25036621	-1.5197448	.835601806	-1.5435791	2.15698242	5.60818481	3.19423675
-2.0721435	2.25036621	-1.5197448	.835601806	-1.5435791	2.15698242	5.60818481	3.19423675	.215454101	-3.1335449	7.32619476	3.81803894
-1.5435791	2.15698242	5.60818481	3.19423675	.215454101	-3.1335449	7.32619476	3.81803894	3.68041992	2.65625	-.15766906	.416818618
.215454101	-3.1335449	7.32619476	3.81803894	3.68041992	2.65625	-.15766906	.416818618	-1.0015869	1.01074218	-.07610130	2.78274345
3.68041992	2.65625	-.15766906	.416818618	-1.0015869	1.01074218	-.07610130	2.78274345	.789794921	2.12341308	4.99722099	4.10923337
-1.0015869	1.01074218	-.07610130	2.78274345	.789794921	2.12341308	4.99722099	4.10923337	1.62231445	-.90576171	5.86753416	4.99679946
.737794921	2.12341308	4.99722099	4.10923337	1.62231445	-.90576171	5.86753416	4.99679946	.756225585	-3.9019775	2.28757095	3.65253555
1.62231445	-.90576171	5.86753416	4.99679946	.756225585	-3.9019775	2.28757095	3.65253555	.668334960	1.93237304	-1.7860630	1.06062328
.756225585	-3.9019775	2.28757095	3.65253555	.668334960	1.93237304	-1.7860630	1.06062328	-3.6962890	-.83923339	2.71113598	3.37606605
.668334960	1.93237304	-1.7860630	1.06062328	-3.6962890	-.83923339	2.71113598	3.37606605	-1.1633300	3.15002441	6.06527504	2.72370320
(1)	.011075283	1.35300715	.936099817	-.12449179	-1.0185055	-.05060984	.748640123	-.47595682	.516893312	-.34907472	.191931866
0	(1)	-.11584441	-.85060870	.483179727	-.38352911	.349391294	-.28033027	.406834247	-.55036669	-1.2237449	-.74483114
0	0	(1)	.816312486	-.12922910	-.48711834	.816312486	.702406092	.259325673	.585097582	.247093185	.293438735
0	0	0	(1)	-1.3530209	-1.0946588	1	1.18068784	.300931214	1.05179191	1.22011814	.345114503
0	0	0	0	(1)	.601895615	-3.852E-17	.013988838	-.47087276	-1.0094892	-.26373642	.449304607
0	0	0	0	0	(1)	5.6296E-17	.073076148	.459392393	.525217644	1.27307614	.846152297
0	0	0	0	0	0	1.9968E-16	(1)	3.16991386	20.7086383	1	2
0	0	0	0	0	0	-1.230E-16	0		(1) .861837333	7.3103E-16	7.2325E-16

0	0	0	0	0	0	6.0083E-16	0	0	1	9.0334E-16	1.1318E-15
0	0	0	0	0	0	-9.397E-17	0	0	0	-4.398E-15	-3.343E-15
0	0	0	0	0	0	3.9045E-16	0	0	0	2.7550E-17	3.6046E-16
0	0	0	0	0	0	1.2750E-16	0	0	0	7.8654E-17	1.3738E-15
0	0	0	0	0	0	6.1942E-17	0	0	0	1.7559E-15	3.3435E-15
0	0	0	0	0	0	7.3875E-16	0	0	0	2.3102E-15	-6.059E-16
1	0	0	0	0	0	-1	0	0	0	4.6464E-16	.25
0	1	0	0	0	0	1.2	0	0	0	1.5047E-14	-.95
0	0	1	0	0	0	1.8127E-16	0	0	0	-6.040E-15	.25
0	0	0	1	0	0	1	0	0	0	2.4438E-14	-1.25
0	0	0	0	1	0	-7.240E-17	0	0	0	-1	2.9024E-16
0	0	0	0	0	1	5.6296E-17	0	0	0	1.2	.7
0	0	0	0	0	0	1.9968E-16	1	0	0	1	2
0	0	0	0	0	0	-1.230E-16	0	1	0	-4.749E-17	-2.522E-16
0	0	0	0	0	0	6.0083E-16	0	0	1	9.0334E-16	1.1318E-15
0	0	0	0	0	0	-9.397E-17	0	0	0	-4.398E-15	-3.343E-15
0	0	0	0	0	0	3.9045E-16	0	0	0	2.7550E-17	3.6046E-16
0	0	0	0	0	0	1.2750E-16	0	0	0	7.8654E-17	1.3738E-15
0	0	0	0	0	0	6.1942E-17	0	0	0	1.7559E-15	3.3435E-15
0	0	0	0	0	0	7.3875E-16	0	0	0	2.3102E-15	-6.059E-16

[a] Primary dependent rows.

[b] Nonprimary dependent row.

Because of $M = [-N_0 \vdots D_0 | -N_1 \vdots D_1 | -N_2 \vdots D_2]$, we have

$$D(z) = \begin{bmatrix} 0 & -1 \\ -0.25 & 1.25 \end{bmatrix} + \begin{bmatrix} 1 & 0 \\ 0 & -2 \end{bmatrix} z + \begin{bmatrix} 0 & 0 \\ 0 & 1 \end{bmatrix} z^2$$

$$= \begin{bmatrix} z & -1 \\ -0.25 & z^2 - 2z + 1.25 \end{bmatrix} \tag{19a}$$

and

$$N(z) = \begin{bmatrix} -1 & 1.2 \\ 0.25 & -0.95 \end{bmatrix} + \begin{bmatrix} 0 & 0 \\ 0 & 0.7 \end{bmatrix} z + \begin{bmatrix} 0 & 0 \\ 0 & 0 \end{bmatrix} z^2$$

$$= \begin{bmatrix} -1 & 1.2 \\ 0.25 & 0.7z - 0.95 \end{bmatrix}. \tag{19b}$$

Thus the transfer matrix of the system is

$$\hat{G}(z) = D^{-1}(z)N(z) = \begin{bmatrix} \dfrac{-z+1}{(z-0.5)^2} & \dfrac{1.2z - 1.1}{(z-0.5)(z-1)} \\ \dfrac{0.25}{(z-0.5)^2} & \dfrac{0.7z - 0.6}{(z-0.5)(z-1)} \end{bmatrix}. \tag{20}$$

This completes the identification. Note that the row degrees of the system are $\nu_1 = 1$ and $\nu_2 = 2$. The degree of the system is $\nu_1 + \nu_2 = 3$. It can be readily shown that the transfer matrix of (17) is indeed equal to the one in (19) and (20).

If a state variable description of the system is desired, we use the procedure in [7, p. 285] to write

$$D(z) = \begin{bmatrix} z & 0 \\ 0 & z^2 \end{bmatrix} \begin{bmatrix} 1 & 0 \\ 0 & 1 \end{bmatrix} + \begin{bmatrix} 1 & 0 & 0 \\ 0 & 1 & z \end{bmatrix} \begin{bmatrix} 0 & -1 \\ -0.25 & 1.25 \\ 0 & -2 \end{bmatrix},$$

$$N(z) = \begin{bmatrix} 1 & 0 & 0 \\ 0 & 1 & z \end{bmatrix} \begin{bmatrix} -1 & 1.2 \\ 0.25 & -0.95 \\ 0 & 0.7 \end{bmatrix}.$$

Thus the observable-form state-variable description of the system is

$$x(n+1) = \left[\begin{array}{c|cc} 0 & 0 & 1 \\ \hline 0.25 & 0 & -1.25 \\ 0 & 1 & 2 \end{array}\right] x(n) + \begin{bmatrix} -1 & 1.2 \\ 0.25 & -0.95 \\ 0 & 0.7 \end{bmatrix} u(n), \tag{21a}$$

$$y(n) = \left[\begin{array}{c|cc} 1 & 0 & 0 \\ 0 & 0 & 1 \end{array}\right] x(n). \tag{21b}$$

This equation is equivalent to the one in (17).

We remark that a number of random sequences are generated to drive the system. All results are surprisingly identical. The obtained transfer matrix is also exact; no numerical error is introduced in the computation. To test the accuracy of the identification procedure, we modify (17) as

$$x(n+1) = \begin{bmatrix} 0.501 & 1 & 0 \\ 0 & 0.5 & 0 \\ 0 & 0 & 1 \end{bmatrix} x(n) + \begin{bmatrix} 0 & 2 \\ 1 & 0 \\ 0 & 0.2 \end{bmatrix} u(n),$$

$$y = \begin{bmatrix} 0.5 & -1 & 1 \\ 0.25 & 0 & 1 \end{bmatrix} x(n).$$

The following polynomial fractional description

$$G(z) = \begin{bmatrix} 0.25 + 124.5z & -125.75z \\ 0.2505 + 123.749z + z^2 & -124.9995 \end{bmatrix}^{-1}$$

$$\times \begin{bmatrix} -124.5 & 150.1 \\ -123.749 - z & 149.1998 + 1.2z \end{bmatrix}$$

is obtained. The result is again exact. This confirms that the algorithm used is a numerically stable one. More simulations for systems with higher degrees are under way.

The matrix $S_\nu(n_1, n_2)$ has a special structure. Each block is obtained by shifting its previous block. By using this shifting property, a fast algorithm may be developed[2].

VI. IDENTIFICATION OF CONTINUOUS-TIME SYSTEMS

We discuss in this section the identification of linear time-invariant lumped continuous-time systems. It will consist of two steps: First use the sampled input-output data to find a discrete-time transfer matrix or state-variable equation and then find a continuous-time transfer matrix or state-variable equation for the system. Consider a continuous-time system with the proper transfer matrix $\hat{G}(s)$ or the state-variable equation

$$\dot{x}(t) = Ax(t) + Bu(t), \qquad y(t) = Cx(t) + Eu(t). \tag{22}$$

If the input u(t) is stepwise during the sampling period, that is,

$$u(t) = u(nT) =: u(n) \quad \text{for} \quad nT \le t < (n + 1)T, \tag{23}$$

where T is the sampling period, then $\{u(n)\}$ and $\{y(n)\}$ can be related by the sampled or discrete-time transfer matrix [13]

$$\hat{G}_d(z) = z\left[\frac{1 - e^{-sT}}{s}\hat{G}(s)\right] = (1 - z^{-1})z[\hat{G}(s)/s], \tag{24}$$

where $z := e^{sT}$, or the discrete-time state-variable equation

$$\begin{aligned} x(n + 1) &= \tilde{A}x(n) + \tilde{B}u(n), \\ y(n) &= \tilde{C}x(n) + \tilde{E}u(n) \end{aligned} \tag{25}$$

with

$$\tilde{A} = e^{AT}, \tag{26a}$$

$$\tilde{B} = \left(\int_0^T e^{At}\, dt\right)B \quad \text{for A singular or nonsingular,}$$
$$= A^{-1}(e^{AT} - I)B \quad \text{for A nonsingular,} \tag{26b}$$
$$\tilde{C} = C \quad \text{and} \quad \tilde{E} = E.$$

If u(t) is stepwise as in (23), then the relationships in (24) and (25) are exact. If u(t) is not stepwise, the relationships are not exact. Clearly the smaller the sampling period, the closer the approximation.

With the preceding discussion, we are ready to identify a continuous-time system from its input and output data. If all poles of the transfer matrix of the continuous-time system are known, the sampling period T should be smaller than π/ω_l, where ω_l is the largest imaginary part of all the poles. Otherwise, the aliasing will occur, or the poles outside the primary strip will fold into the primary strip, and it would be impossible to obtain a correct continuous-time transfer matrix from the identified discrete-time transfer matrix [13]. In practice, the poles are not known. In this case, we may choose the sampling period T as follows. Find the smallest distance P between nearby peak and valley and then set T = 0.8P [14]. For a much deeper discussion of the sampling period, see [3,4,15].

Once a sampling period T is chosen, we may then use $\{u(n) := u(nT)\}$ and $\{y(n) := y(nT)\}$ to find a sampled transfer matrix $\hat{G}_d(z) = D^{-1}(z)N(z)$. Once $\hat{G}_d(z)$ is found, we use (24) to compute

$$\hat{G}(s) = sz^{-1}\left[\frac{z}{z-1}\hat{G}_d(z)\right]. \tag{27}$$

This will be an approximation of the transfer matrix of the continuous-time system. If the input u(t) happens to be stepwise, as in the case of digital computer control, then the $\hat{G}(s)$ in (27) will be an exact description of the system.

Once $\hat{G}_d(z) = D^{-1}(z)N(z)$ is identified, a discrete-time state-variable equation can also be readily obtained, as shown in the example in Section V. To find a continuous-time equation, we must solve A from $\tilde{A} = e^{AT}$. We remark that if $\tilde{A}$ is obtained from a continuous-time system, it is nonsingular or, equivalently, its eigenvalues are all nonzero. From $\tilde{A} = e^{AT}$, we have

$$A = [\ln \tilde{A}]/T. \tag{28}$$

This can be computed using a polynomial f(x) that is equal to ln x on the spectrum of $\tilde{A}$ or by transforming $\tilde{A}$ into a Jordan form [7]. A recursive equation is also suggested in [16] to compute (28). The recursive equation, however, may not converge. Thus a convergent recursive algorithm for computing A from $\tilde{A}$ is still lacking.

Once A is computed, B, C and E can be computed from (26b). The computed {A, B, C, E} is an approximate description of the continuous-time system if the input is not stepwise. Otherwise it is an exact description.

VII. NOISY MEASUREMENTS

In practice, measurements are often corrupted with noises. Identification using noisy data has been extensively studied. A number of recursive algorithms are available in [2,17]. In recursive identifications, the row degrees are required to be known.

In this section, we suggest a method to identify a system where the output signals are corrupted by noises. The method must have been used in practice and no originality is claimed.

The purpose is to apply the method in this paper to noisy data and then compare with existing methods.

If the output signals are corrupted by noises as $y(n) = s(n) + v(n)$, where $\{s(n)\}$ is due to input and $\{v(n)\}$ is noise, we pass each output component $y_i(n)$ through a filter with a known transfer function $h_i(z)$. The output of each filter is denoted by $\tilde{y}_i$. We then use $\{u(n)\}$ and $\{\tilde{y}(n)\}$ to find a transfer matrix $\hat{G}_1(z)$. If the transfer matrix of the original system is $\hat{G}(z)$, then we have

$$\hat{G}_1(z) = \text{diag}\{h_1(z), h_2(z), \ldots, h_q(z)\}\hat{G}(z).$$

Thus the transfer matrix of the system is

$$\hat{G}(z) = \text{diag}\left\{h_1^{-1}(z), h_2^{-1}(z), \ldots, h_q^{-1}(z)\right\}\hat{G}_1(z).$$

If $h_1(z) = h_2(z) = \cdots = h_q(z) = h(z)$, then

$$\hat{G}(z) = h^{-1}(z)\hat{G}_1(z).$$

This approach will be extensively tested on a digital computer. The results will also be compared with the ones obtained by recursive algorithms.

ACKNOWLEDGMENT

The author wishes to thank Stanley Chen for carrying out the computer computation.

REFERENCES

1. R. KALABA and K. SPRINGARN, "Control Identification and Input Optimization," Plenum, New York, 1982.

2. L. LJUNG and T. SODERSTROM, "Theory and Practice of Recursive Identification," MIT Press, Cambridge, Massachusetts, 1983.

3. G. GOODWIN and R. L. PAYNE, "Dynamic System Identification," Academic Press, New York, 1977.

4. R. K. MEHRA and D. G. LAINIOTIS, "System Identification: Advances and Case Studies," Academic Press, New York, 1976.

5. D. GRAUPE, "Identification and Adaptive Filtering," Krieger, Malabar, Florida, 1984.

6. T. C. HSIA, "Identification: Least Square Methods," Lexington Books, Lexington, Massachusetts, 1977.

7. C. T. Chen, "Linear System Theory and Design," Holt, New York, 1984.

8. E. NERING, "Linear Algebra and Matrix Theory," 2nd ed., Wiley, New York, 1970.

9. G. W. STEWART, "Introduction to Matrix Computations," Academic Press, New York, 1973.

10. K. J. ASTROM and T. BOHLIN, "Numerical Identification of Linear Dynamic Systems from Normal Operating Records," *in* "Theory of Self-Adaptive Control Systems," (P. H. Hammond, ed.), Plenum, New York, 1966.

11. B. D. O. ANDERSON and C. R. JOHNSON, JR., "Exponential Convergence of Adaptive Identification and Control Algorithms," *Automatica 18*, 1-13 (1982).

12. J. S. C. YUAN and W. M. WONHAM, "Probing Signals for Model Reference Identification," *IEEE Trans. Autom. Control Ac-22*, 530-536 (1977).

13. B. C. KUO, "Digital Control Systems," Holt, New York, 1980.

14. C. T. CHEN, "One-Dimensional Digital Signal Processing," Dekker, New York, 1979.

15. P. DATZ, "Digital Control Using Microprocessors," Prentice-Hall, Englewood Cliffs, New Jersey, 1981.

16. N. K. SINHA and G. J. LASTMAN, "Identification of Continuous-Time Multivariable Systems from Sampled Data," *Int. J. Control 35*, 117-126 (1982).

17. A. NEHORAI and M. MORF, "Recursive Identification Algorithms for Right Matrix Fraction Description Models," *IEEE Trans. Autom. Control AC-29*, 1103-1106 (1984).

18. C. T. CHEN, "Adaptive Control of Linear Multivariable Systems without any Prior Knowledge," *Proc. IEEE Conf. Decision Control*, 1882-1887 (1985).

19. J. J. DONGARRA, C. B. MOLER, J. R. BUNCH, and G. W. STEWART, "LINPACK User's Guide," *SIAM*, Philadelphia, Pennsylvania, 1979.

Techniques for the Selection of Identifiable Parametrizations for Multivariable Linear Systems

MICHEL GEVERS
VINCENT WERTZ

Laboratoire d'Automatique,
de Dynamique et d'Analyse des Systèmes
Louvain University
B-1348 Louvain la Neuve, Belgium

I. INTRODUCTION

This article is a survey. The objective is to give an up-to-date account of the use of identifiable representations for linear multivariable systems as well as to briefly survey methods for the estimation of the structure of these representations from measured data. Linear multivariable systems can be represented in a number of ways using finitely parametrized models; in this article we will consider state-space models (SSs), matrix fraction descriptions (MFDs), and autoregressive moving average models with exogeneous inputs (ARMAXs), also called vector difference equation models (VDEs). This last class of models is of particular importance in the context of system identification from observed data because most available parameter estimation methods are well suited to ARMAX models. Transformations between the various classes of finite-dimensional models are easy to establish, and within each class there exists an infinite number of equivalent models, all producing the same sequence of Markov parameters (or impulse response matrices $\{K_0, K_1, K_2, \ldots\}$ or, equivalently, the same matrix transfer function $K(z) \triangleq \Sigma_{i=0}^{\infty} K_i z^{-i}$.

A system is uniquely defined by its sequence of Markov parameters. If the system is finite dimensional (see Section II), it can be represented in an infinite number of ways by a finitely parametrized model (SS, MFD, or ARMAX). Obviously, in an identification context, one would like to use a parametrized model set $M^* = \{M(\theta) | \theta \in D\}$ (either an SS, MFD, or ARMAX model set) that is able to represent the system for a unique value of the parameter vector θ, so that the parameter estimation algorithm would converge. This is the problem of constructing identifiable model sets to which this article is devoted. The structure of these model sets is determined by a set of integer-valued indices, called structure indices. They determine the locations of 0 and 1 elements and of free parameters in the matrices of an SS model set or the degrees of the polynomials in an MFD or ARMAX model set.

A fundamental property of linear multivariable systems is that no unique model set is able to represent all systems of given order, say n. The set S(n) of all systems of order n can only be represented as a finite union of model sets, each characterized by its structure indices. This is what makes the determination of identifiable model sets nontrivial in the multivariable case. Now there are basically two ways to let each system in S(n) be represented by an identifiable model. One way is to let S(n) be described by a disjoint union of identifiable model sets: that is, each system $\sigma \in S(n)$ is represented by a unique model specified by a unique set of integer-valued structure indices and a unique set of real-valued parameters. These identifiable model sets are then called canonical. Given a particular choice of canonical form (e.g., observer canonical form), there exists one such form for each system. The alternative is to let S(n) be described by a union of identifiable but overlapping model sets, each set being characterized by its set of integer-valued structure indices and each model within that set by a unique set of real-valued parameters. In this approach the structure indices are not defined by the system. As a matter of fact, each of the model sets is dense in S(n) (see [1]-[4] for a description of the structure of S(n)); this means that any such identifiable model set M^* will be able to represent almost any given system $\sigma \in S(n)$ for a particular value of θ. If $M_1(\theta_1)$ and $M_2(\theta_2)$ are the representations of a same system σ in two different identifiable model sets, then the parameter vectors θ_1 and θ_2 are related by a transformation that corresponds to a coordinate transformation in Euclidean space.

Canonical forms for multivariable systems were first introduced as a tool to simplify some observer or controller design problems (see, e.g., [5]). Because a canonical form is also a uniquely defined representative of an equivalence class of (SS, MFD, or ARMAX) models, its importance in identification was soon recognized (see [1] and [6]-[12]). When using canonical forms for system identification, the most critical part of the problem is the estimation of the structure indices (or Kronecker indices) of the system. If they have been wrongly estimated, then the parameter vector θ cannot converge to the true θ_0 [13].

The estimation of the structure indices becomes very critical if the system happens to lie close to the boundary between two of the disjoint subsets of S(n) mentioned earlier.

Around 1974, the structure of S(n) came to be much better understood thanks to the important work of Glover and Willems [8], Hazewinkel and Kalman [14], and Clark [2], who showed that S(n) has the structure of an analytic manifold that can be covered by a union of overlapping subsets and that an identifiable parametrization can be defined for each subset. These parametrizations were called overlapping forms or pseudocanonical forms and their use for identification of multivariable linear systems, as an alternative to canonical forms, was studied by a number of authors: overlapping state-space models were examined in [15] through [18], ARMAX models in [19] through [20], while both state-space and ARMAX models, and the relationships between them, were studied in [21] through [26]. The advantage of overlapping forms over canonical forms is that, instead of having to estimate p structure indices $n_1, \ldots, n_p$ (p being the number of observed outputs of the system), one has to estimate only the order of the system. Given the order n, one can choose any set of structure indices $n_1, \ldots, n_p$ adding up to n, and, almost surely, one can identify the system in the corresponding pseudocanonical (or overlapping) form. Numerical considerations may lead one to choose a particular set of indices (see [16]-[17]). The disadvantage is that this form may contain a few parameters more than the canonical form.

In any case, whether canonical or pseudocanonical forms are used, one has to estimate one or more integer-valued indices from the data: the order n or the structure indices $n_1, \ldots, n_p$. This is called structure estimation; it is the most critical step in the identification of multivariable systems. (In econometrics, structure estimation is actually called identification.) Recent work of Hannan and Kavalieris [26]-[28] has shown that, under mild conditions, the order n (for pseudocanonical forms) and the Kronecker indices $n_1, \ldots, n_p$ (for canonical forms) can be consistently estimated. See also [25] for an excellent presentation of these recent results and for a discussion of the consequences of misspecifying the structure.

Here we will survey most of the results mentioned. Because almost all results are available in the literature, we shall give them without proof, but refer to the appropriate references. In Section II we shall present the different model sets discussed here and establish their interconnection. The concept of identifiability will be presented in Section III, while the structure of S(n) will be analyzed in Section IV. It will be shown that a rational system can be represented by a point (i.e., a coordinate vector) in an appropriate coordinate system. In Section V we will present a class of canonical and pseudocanonical model structures, in SS, MFD, and ARMAX form; these structures are all derived from the coordinates of the system, defined in a more abstract way in Section IV. In Section VI we will briefly present some other identifiable model structures; they are not directly derived from the coordinate vector defined in Section IV and tend to have a larger number of parameters. Finally, in Section VII we will survey the most recent results on structure estimation; this means estimation of the Kronecker indices when canonical forms are used, of the order of the system when pseudocanonical forms are used.

II. MODELS

As a starting point, we consider a p-vector stationary stochastic process y(t) generated as follows:

$$y(t) = G(z)u(t) + H(z)e(t), \tag{1}$$

where $y(t) \in R^p$, $u(t) \in R^m$, $e(t) \in R^p$, and G(z) and H(z) are causal rational transfer function matrices. The following assumptions are made about the system:

(i) $E\{e(t)\} = 0, \quad E\{e(t)e^T(s)\} = \Sigma\delta_{ts}, \quad \Sigma > 0,$ (2a)

(ii) $G(z) = G_1 z^{-1} + G_2 z^{-2} + \cdots,$ (2b)

$H(z) = I + H_1 z^{-1} + H_2 z^{-2} + \cdots,$ (2c)

(iii) G(z) is analytic in $|z| \geq 1$, (2d)

(iv) H(z) has full rank so that $H^{-1}(z)$ exists and H(z) and $H^{-1}(z)$ are analytic in $|z| \geq 1$, and (2e)

(v) u(t) is an observed input signal, which can be either deterministic or stochastic, but we assume that $|u(t)|$ is bounded and that the following limits exist:

$$\lim_{N\to\infty} \frac{1}{N} \sum_{1}^{N} E[u(t)] = 0,$$
$$\lim_{N\to\infty} \frac{1}{N} \sum_{1}^{N} E[u(t)u^T(t-\tau)] = R_u(\tau), \tag{2f}$$

where the expectation is discarded if u(t) is deterministic. We then define the spectrum of $u(\cdot)$ as

$$\phi_u(\omega) \triangleq \sum_{\tau=-\infty}^{\infty} R_u(\tau)e^{i\omega\tau}. \tag{3}$$

The model (1)-(2) is very general and can be justified as follows. If

$$y(t) = s(t) + v(t) = G(z)u(t) + v(t), \tag{4}$$

where s(t) is some useful signal and v(t) is noise, and if the spectrum $\phi_v(\omega)$ of v(t) is rational, Hermitian, and positive definite for all ω in $[-\pi, \pi]$, then $\phi_v(\omega)$ can be decomposed uniquely as

$$\phi_v(\omega) = H(e^{i\omega})\Sigma H^T(e^{-i\omega}), \tag{5}$$

where H(z) and Σ satisfy the conditions (2a, c, e).

With these conditions, the e(t) are the linear innovations, that is, the prediction errors of the best linear one-step-ahead prediction of v(t) from its infinite past. The model (1) will be called a transfer function (TF) model.

Comment 1. The condition (2d) has been introduced to make y(t) a stationary process, so that covariances and spectra can be defined. It is necessary for most of the consistency results on structure estimation, which will be presented in Section VII. However, it is not required for the proper definition of canonical and pseudocanonical forms; obviously one should be able to identify unstable dynamical systems. The important feature for this is that the predictor $\hat{y}(t|t-1)$ be stable, rather than the data-generating model; see Comment 3 later.

With the model (1)-(2) we shall associate a Hankel matrix $H_{1,\infty}[G \vdots H]$ defined as follows. Let $\{K_1, K_2, \ldots\}$ be a sequence of matrices; then

$$H_{1,N}[K] \triangleq \begin{bmatrix} K_1 & K_2 & \cdots & K_N \\ K_2 & K_3 & \cdots & K_{N+1} \\ \vdots & & & \\ K_N & K_{N+1} & \cdots & K_{2N-1} \end{bmatrix} \tag{6}$$

In particular $H_{1,\infty}[G \vdots H]$ is obtained by replacing K_i in (6) by $[G_i \vdots H_i]$ and by letting N go to ∞. The rank of $H_{1,\infty}[G \vdots H]$ is then called the order of the system (1).

Given the spectra $\phi_y(\omega)$, $\phi_u(\omega)$, and $\phi_{yu}(\omega)$ (all of which can theoretically be estimated from arbitrarily long input and output records), in principle one can determine G(z) and H(z) uniquely under weak conditions on $\phi_u(\omega)$ from

$$\phi_{yu}(\omega) = G(e^{i\omega})\phi_u(\omega), \tag{7}$$

$$\phi_y(\omega) = G(e^{i\omega})\phi_u(\omega)G^T(e^{-i\omega}) + H(e^{i\omega})\Sigma H^T(e^{-i\omega}). \tag{8}$$

However, to actually construct an estimate of G(z) and H(z) from second-order statistics, one needs to represent the process y(t) in a finitely parametrized form and then construct an algorithm for the estimation of these parameters. In other words, one needs to define a coordinate space and to let the system be represented by a point in that coordinate space. This is the problem of parametrization, that is, of defining finite-dimensional identifiable model sets for the process y(t). From now on we shall consider (1)-(2) as the given system, and we shall study three classes of finitely parametrized model sets that are input-output equivalent with the system (1)-(2). Notice that (1)-(2) is an infinite-dimensional representation; that is, the system is specified by the infinite sequence of matrices $[G_1, H_1, G_2, H_2, \ldots]$.

A. STATE-SPACE MODELS

One way of representing the system (1)-(2) is through a state-space model:

$$\begin{cases} x(t+1) = Ax(t) + Bu(t) + Ke(t), & (9a) \\ y(t) = Cx(t) + e(t), & (9b) \end{cases}$$

where y(t), u(t), and e(t) are the same as in (1), and where dim $x(t) \triangleq n$ is minimal, with A, B, C, K such that

$$C(zI - A)^{-1}[B \,\vdots\, K] = [G(z) \,\vdots\, H(z) - I]. \tag{10}$$

By conditions (2d, e) this implies that all the eigenvalues of A have modulus less than 1. It is a standard result of linear system theory that

$$n \triangleq \dim x(t) = \operatorname{rank} H_{1,\infty}[G \vdots H], \tag{11}$$

where n is called the order of the system. Now there exists an infinity of SS models {A, B, C, K} that are input-output equivalent with the system (1)-(2), that is, for which the relation (10) holds. They are all obtained from an arbitrary model by the following similarity transformations

$$A^* = T^{-1}AT, \quad B^* = T^{-1}B, \quad K^* = T^{-1}K, \quad C^* = CT, \tag{12}$$

where T is any nonsingular $n \times n$ matrix. Our task is to define identifiable model sets, that is, to parametrize $A(\theta)$, $B(\theta)$, $C(\theta)$, $K(\theta)$ in such a way that any arbitrary system $\sigma \in S(n)$ (defined by the sequences G_i, H_i; $i = 1, 2, \ldots$) can be represented in this parametrized model set by a unique value of the parameter vector θ.

B. *MATRIX FRACTION DESCRIPTIONS*

Another representation of the system (1)-(2) is

$$P(z)y(t) = Q(z)u(t) + R(z)e(t), \tag{13}$$

where P(z), Q(z), and R(z) are left-coprime polynomial matrices (see, e.g., [29]) satisfying the relationship

$$P^{-1}(z)[Q(z) \vdots R(z)] = [G(z) \vdots H(z)]. \tag{14}$$

In (13) z is the forward shift operator: $zy(t) \triangleq y(t+1)$. If n is the order of the system (see above), then

$$\deg \det P(z) = n. \tag{15}$$

Also, by (2d, e), det P(z) and det R(z) have all their roots in $|z| < 1$.

Again the MFD (13) is nonunique. If $\{P(z), Q(z), R(z)\}$ is an arbitrary left-coprime MFD for (1)-(2), then all input-output-equivalent left-coprime MFD models are obtained from this arbitrary model by the following unimodular transformations:

$$[P^*(z) \vdots Q^*(z) \vdots R^*(z)] = U(z)[P(z) \vdots Q(z) \vdots R(z)], \tag{16}$$

where $U(z)$ ranges over the set of all unimodular matrices of dimension $p \times p$. (A unimodular matrix is a square polynomial matrix whose determinant is a nonzero constant.) Again, to obtain identifiable MFD model sets, our task will be to parametrize $P(z, \theta)$, $Q(z, \theta)$, $R(z, \theta)$ (i.e., to define their structure) in such a way that (14) holds for a unique value of θ.

C. VECTOR DIFFERENCE EQUATION OR ARMAX MODELS

One of the most widely used model sets in econometrics, but also in engineering, is the ARMAX[1] model set

$$\bar{P}(D)y(t) = \bar{Q}(D)u(t) + \bar{R}(D)e(t), \tag{17}$$

where $\bar{P}(D)$, $\bar{Q}(D)$, and $\bar{R}(D)$ are left-coprime polynomial matrices with

$$\bar{P}(D) = \bar{P}_0 + \bar{P}_1 D + \cdots + \bar{P}_p D^p, \tag{18a}$$

$$\bar{Q}(D) = \bar{Q}_1 D + \cdots + \bar{Q}_q D^q, \tag{18b}$$

$$\bar{R}(D) = \bar{R}_0 + \bar{R}_1 D + \cdots + \bar{R}_r D^r, \tag{18c}$$

where D is the delay operator: $Dy(t) \triangleq y(t-1)$.

In order for (17)-(18) to be a representation of (1)-(2), the following conditions must hold:

$$\bar{P}^{-1}(D)[\bar{Q}(D) \vdots \bar{R}(D)] = [\bar{G}(D) \vdots \bar{H}(D)], \tag{19}$$

[1] *When there is no exogenous input $u(t)$, Eq. (17) becomes $\bar{P}(D)y(t) = \bar{R}(D)e(t)$ and is called an ARMA model.*

where

$$\bar{G}(D) \triangleq G(D^{-1}) = \sum_{1}^{\infty} G_i D^i, \tag{20}$$

and similarly for $\bar{H}(D)$. In particular (19) implies that

(i) $\bar{P}_0$ is nonsingular and $\bar{P}_0 = \bar{R}_0$, (21)

(ii) det $\bar{P}(z)$ and det $\bar{R}(z)$ have all their roots in $|z| > 1$. (22)

In addition to (21), we will often want ARMAX models for which

$$\bar{P}_0 = \bar{R}_0 = I_p. \tag{23}$$

This allows one to write

$$y(t) = -\sum_{1}^{p} \bar{P}_i y(t - i) + \sum_{1}^{q} \bar{Q}_i u(t - i)$$

$$+ \sum_{1}^{r} \bar{R}_i e(t - i) + e(t). \tag{24}$$

We shall see later that condition (23) introduces some additional complications.

Comment 2. We could have used z^{-1} for the delay operator in lieu of D. We have not done so because we want to keep powers of z^{-1} for power series and to stress that $\bar{P}(D)$, $\bar{Q}(D)$, and $\bar{R}(D)$ are polynomials.

There is an obvious relationship between MFD models and ARMAX models. Let (14) be a left-coprime MFD of $[G(z) \vdots H(z)]$ and let $n_1, \ldots, n_p$ be the row degrees of the matrix $[P(z) \vdots Q(z) \vdots R(z)]$.

Define $M(z) = \text{diag}\{z^{n_1}, \ldots, z^{n_p}\}$, and let $z^k D^k = D^k z^k = 1$ so that $M(z)M(D) = M(D)M(z) = I_p$. Then

$$M(D)[P(z) \vdots Q(z) \vdots R(z)] = [\bar{P}(D) \vdots \bar{Q}(D) \vdots \bar{R}(D)]. \tag{25a}$$

Similarly

$$M(z)[\bar{P}(D) \vdots \bar{Q}(D) \vdots \bar{R}(D)] = [P(z) \vdots Q(z) \vdots R(z)]. \tag{25b}$$

Note that the row degrees of $[P(z) \vdots Q(z) \vdots R(z)]$ are identical to the row degrees of $[\bar{P}(D) \vdots \bar{Q}(D) \vdots \bar{R}(D)]$; however, the row degrees of $\bar{P}(D)$ are not equal to those of $P(z)$, and deg det $\bar{P}(D)$ is not equal to the order of the system (see [30] and [31] for details).

Comment 3. An alternative to using the "data-generating model" (1)-(2) as our starting point is to use a "predictor model," that is, the model that generates the one-step-ahead predictions of $y(t)$ given the infinite past. It can be written

$$\hat{y}(t|t-1) = W_u(z)u(t) + W_y(z)y(t), \tag{26}$$

where $W_u(z)$ and $W_y(z)$ are stable rational transfer function matrices

$$W_u(z) = \sum_1^{\infty} W_u(k)z^{-k}, \quad W_y(z) = \sum_1^{\infty} W_y(k)z^{-k}. \tag{27}$$

The model (26) is easily derived from the data-generating model (1)-(2):

$$W_u(z) = H^{-1}(z)G(z), \quad W_y(z) = I - H^{-1}(z). \tag{28}$$

The use of the prediction model (26) as a starting point for the analysis of identifiable model sets is justified by the fact that most identification methods are based on minimizing prediction errors. This viewpoint is taken in [32]. All our subsequent analysis for the representation of (1) by finitely parametrized identifiable model sets applies equally well to the representation (26), which can also be modeled by SS, MFD, or ARMAX models. In SS form, (26) leads to the Kalman filter. Notice finally that the condition "$\{W_u(z), W_y(z)\}$ stable" is really what is required to apply prediction error methods; it enables one to identify some models of the form (1) with unstable $G(z)$. For example, an MFD model (13) with $\det P(z)$ having roots in $|z| \geq 1$ and $\det R(z)$ having all its roots in $|z| < 1$ leads to an unstable $G(z)$, but yields a stable $W_u(z)$ and $W_y(z)$.

III. IDENTIFIABILITY

There are several ways of defining identifiability. Here we propose a setup and a definition inspired by, but not identical to, [32]. As a starting point we assume that we have an input process $u(t)$ and an output process $y(t)$ that can be described by (1):

$$y(t) = G(z)u(t) + v(t) = G(z)u(t) + H(z)e(t) \tag{29}$$

with the properties (2). This is our basic model. By the constraints (2) $H(z)$ is a unique factorization of $\phi_v(\omega)$. Therefore $G(z)$, $H(z)$, and Σ are uniquely defined by the spectra ϕ_y, ϕ_{uy}, and ϕ_u (see Section II), and therefore (29) can actually be seen as an input/output (I/O) model from the input process

$$\begin{bmatrix} u \\ e \end{bmatrix}$$

to the output process y with transfer function matrix $[G(z) \vdots H(z)]$. Note that we could also have chosen (26) as our basic model, which would similarly be uniquely defined by the spectra.

Since we also want to consider other classes of models than the TF model, such as SS, MFD, or ARMAX models, we introduce the following definition.

Definition 1. A model M is a stable algebraic operator that transforms a given input process

$$\begin{bmatrix} u(t) \\ e(t) \end{bmatrix}$$

into a unique output process y(t).

Therefore (1), (9), (13), and (17) are all models by our definition, provided that specific values (rational functions or real-valued parameters or polynomials) are put into the operators G(z), H(z), A, B, C, K, ... Whatever the description chosen for our model (SS, MFD, ARMAX), we can always compute the corresponding transfer functions G(z) and H(z) from it [see, e.g., (10)]. We can therefore introduce the following definition.

Definition 2. Let $M^{(1)}$ and $M^{(2)}$ be two models relating an input vector

$$\begin{bmatrix} u \\ e \end{bmatrix}$$

to an output vector y, and let $\{G^{(1)}(z), H^{(1)}(z)\}$ and $\{G^{(2)}(z), H^{(2)}(z)\}$ be the corresponding transfer functions. Then the two models are equivalent [we write $M^{(1)} = M^{(2)}$] if and only if $G^{(1)}(z) = G^{(2)}(z)$ and $H^{(1)}(z) = H^{(2)}(z)$.

In identification a search will typically be conducted over a set of models. Most often this model set is noncountable; it is obtained as the range of a smoothly parametrized algebraic operator where the parameter vector θ is allowed to cover a subset D of R^d ($d = \dim \theta$). The search for the "best" model is then performed over all $\theta \in D$. We formalize this as follows.

Definition 3. A model set M^* is a set of models: $M^* = \{M_\alpha | \alpha \in I_\alpha\}$, where I_α is an index set.

Most often the index set is noncountable. The model set is then smoothly parametrized by a parameter vector θ of dimension d, where $\theta \in D_M \subset R^d$.

Definition 4. A model structure $\tilde{M}$ is a differentiable mapping from a subset D_M of R^d to a model set: $\tilde{M} : \theta \in D_M \to M(\theta) \in M^*$, where $M(\theta)$ is defined by any one of the stable parametrized algebraic operators described earlier.

Example 1. We illustrate by a scalar example.

(1) $M : y(t) = 0.85\, y(t - 1) + 1.5\, u(t - 1)$.

(2) $M^* = \left\{y(t) = \alpha y(t - 1) + \beta u(t - 1) \mid (\alpha, \beta) \in D_M \subset R^2\right\}$, where $D_M = \{\alpha, \beta \mid |\alpha| < 1, |\beta| \leq 100\}$, for example.

(3) $\tilde{M} : (\alpha, \beta) \in D_M \subset R^2 \to M^* = \{y(t) = \alpha y(t - 1) + \beta u(t - 1) \mid (\alpha, \beta) \in D_M\}$.

The range of a model structure defines a model set: Range $\tilde{M} = \{M(\theta) \mid \theta \in D_M\} = M^*$. Note that a given model set can typically be described as the range of different model structures. An important problem, which will be discussed at length in this paper, is to find a model structure (i.e., a parametrization) whose range equals a given model set. It turns out that the set S(n) of all linear systems of order n with dim $y_t = p > 1$ and dim $u_T = m > 1$ cannot be described as the range of a single model structure. The remedy will be to describe S(n) as a union of ranges of different model structures:

$$M^* = \bigcup_{i=1}^{l} R\left(\tilde{M}_i\right). \tag{30}$$

We can now define identifiability. The concept of identifiability involves several aspects and has therefore given rise to several definitions. In its broad sense, identifiability is concerned with whether the identification procedure yields a unique value of θ, and possibly whether the resulting model $M(\theta)$ is equivalent to the true system σ. This involves aspects of whether $\sigma \in M^*$, whether the data set is informative enough, and whether the "model structure is identifiable." This last problem, which we shall call structural identifiability, concerns only the invertibility of the map of Definition 4.

Definition 5. A model structure $\tilde{M}$ is called *globally identifiable at* θ^* if

$$M(\theta) = M(\theta^*), \quad \theta \in D_m \Rightarrow \theta = \theta^*. \tag{31}$$

It is called *strictly globally identifiable* if it is globally identifiable $\forall\theta^* \in D_M$.

This last condition is almost never met, as we shall see in a moment. Therefore we introduce a more realistic property.

Definition 6. A model structure $\tilde{M}$ is *globally identifiable* if it is globally identifiable at almost all $\theta^* \in D_M$.

Similar definitions can be introduced for local identifiability, that is, where (31) holds for θ in a neighborhood of θ^*. We stress that these definitions cover only one aspect of identifiability; in particular they are a property of the model structure only and are totally independent of a possible "true system." To stress the relevance of these definitions to our problem of constructing identifiable model structures, we point out the following facts:

(1) The set of all single-input/single-output models of order up to n cannot be described as the range of a strictly globally identifiable model structure without restrictions on θ. This is because of pole-zero cancellations on certain hypersurfaces. It can, however, be described as the range of a globally identifiable model structure.

(2) The set of all multivariable systems of order up to n (or even exactly n) cannot be described as the range of a unique globally identifiable model structure. We shall see in the next section that it can be described as a union of ranges of globally identifiable model structures. The concept of structural identifiability is very old in econometrics. It was first introducted in the engineering literature in [33]. It is important for numerical reasons. It guarantees that if the data are informative enough, the parameter estimation problem will be well posed, since different θ yield different I/O properties. The definition we have adopted is by no means unique. Instead of using the uniqueness of $G(z, \theta)$ and $H(z, \theta)$, some definitions use the uniqueness of the parametrized joint probability function of the data $p(Y, U; \theta)$ as a starting point (see, e.g., [18], [23], [35]). Other definitions are based on the Kullback-Leibler information or on the theory of estimable functions (see [35,36]). Notice finally that the term "structural identifiability" has a different meaning in econometrics [25].

IV. THE STRUCTURE OF S(n)

Consider $K(z) \triangleq [G(z) \vdots H(z) - I] = \Sigma_1^\infty K_i z^{-i}$ of dimension $p \times s$, with $s \triangleq m + p$. Then $K(z)$ is strictly proper, that is,

$$\lim_{z\to\infty} K(z) = 0. \tag{32}$$

Definition 1. We call S(n) the set of all strictly proper stable rational transfer function matrices K(z) of order n (i.e., such that rank $H_{1,\infty}[K] = n$).

Notice that the problem of representing $[G(z) \vdots H(z)]$ by an identifiable model is a trivial modification of the problem of representing $[G(z) \vdots H(z) - I]$, and this last problem is that of representing an arbitrary strictly proper transfer function $K(z)$. This is why we now study the structure of $S(n)$. If we further denote

$$m(t) \triangleq \begin{bmatrix} u(t) \\ e(t) \end{bmatrix},$$

then the basic model (1) can be written

$$y(t) = K(z)m(t) + e(t), \tag{33}$$

with $m(t) \in R^s$, and the problem becomes one of finding an identifiable representation of $K(z)$. The cases where $u(t) = 0$ (ARMA model) or $e(t) = 0$ (I/O model) are special cases of our setup.

The topology on $S(n)$ is the relative pointwise topology T_{pt} in which a sequence $K^{(n)}(z)$ converges to $K(z)$ if and only if the coefficient matrices $K_i^{(n)}$ converge to K_i (in the relative Euclidean topology) for all $i = 1, 2, \ldots$ (see [3]). If S is a set in this topological space, its closure will be denoted by $\bar{S}$. In particular $\bar{S}(n)$ denotes the set of all strictly proper rational transfer functions of order less than or equal to n.

Now let $K(z)$ be a point in $S(n)$, with $K(z)$ of dimensions $p \times s$, and consider the Hankel matrix $H_{1,\infty}[K]$. This Hankel matrix is made up of blocks of p rows. Call r_{ij} the ith row of the jth block of $H_{1,\infty}[K]$. We shall also make the following standing technical assumption.

Technical Assumption. The rows of the first block of $H_{1,\infty}[K]$ are linearly independent.

This really only eliminates degenerate transfer functions $K(z)$ whose rows would be linearly dependent. We now state without proof a few important results on $S(n)$.

Result 1. If $K(z) \in S(n)$, then there exists at least one partition $n = n_1 + n_2 + \cdots + n_p$ of n such that the set of rows

$$\{r_{ij} : i = 1, \ldots, p;\ j = 1, \ldots, n_i\} \triangleq R(n_1, \ldots, n_p) \tag{34}$$

form a basis of $H_{1,\infty}[K]$.

This follows immediately from the Hankel structure. Note that the rows in (34) are chosen in such a way that if $r_{ij} \in R(n_1, \ldots, n_p)$ for $j > 1$, then $r_{i,j-1} \in R(n_1, \ldots, n_p)$. The set of the row indices corresponding to the rows of $R(n_1, \ldots, n_p)$ is completely determined by the partition $\mu \triangleq (n_1, \ldots, n_p)$. The indices $n_1, \ldots, n_p$ are called structure indices. We also denote $|\mu| = \Sigma_1^p n_i = n$. It is easy to see that there are

$$\begin{bmatrix} n + p - 1 \\ p - 1 \end{bmatrix}$$

such partitions.

Definition 2. We call U_μ the set of all points in $S(n)$ for which the corresponding set of rows (34) specified by μ forms a basis for the rows of $H_{1,\infty}[K]$.

We now show that U_μ can be completely coordinatized by $n(p + s)$ coordinates. Row r_{i,n_i+1} is a unique linear combination of the basis rows

$$r_{i,n_i+1} = \sum_{j=1}^{p} \sum_{l=1}^{n_j} \alpha_{ijl} r_{jl}, \quad i = 1, \ldots, p. \tag{35}$$

It follows again from the Hankel structure that knowing the first block of elements (i.e., the first s elements) of the basis rows (34) and the coefficients $\{\alpha_{ijl};\ i,j = 1, \ldots, p;\ l = 1, \ldots, n_j\}$ allows one to compute any other row of $H_{1,\infty}[K]$

and therefore to specify K(z) completely. Therefore any element $K(z) \in U_\mu$ can be mapped into a vector τ_μ in R^d by the following $d \triangleq n(p + s)$ coordinates:

$$\tau_\mu = \begin{Bmatrix} \alpha_{ijl}; & i, j = 1, \ldots, p; & l = 1, \ldots, n_j & \\ k_l(i, j); & i = 1, \ldots, p; & j = 1, \ldots, s; & l = 1, \ldots, n_i \end{Bmatrix} \tag{36}$$

where $k_l(i, j)$ is the (i, j)th element of the matrix K_l. We shall call ϕ_μ the mapping from K(z) into τ_μ:

$$\Phi_\mu: U_\mu \to \Theta_\mu \subset R^d: K(z) \in U_\mu \to \tau_\mu = \Phi_\mu(K(z)) \in \Theta_\mu. \tag{37}$$

We now have the following important result, originally proved by Clark [2], and further extended by several authors (see, e.g., [14] and [19]).

Result 2.

(1) S(n) is a real analytic manifold of dimension n(p + s).

(2) S(n) is the union of the U_μ such that $|\mu| = n$. Each U_μ is open and dense in S(n); Θ_μ is open and dense in R^d.

(3) Φ_μ described in (37) is a homeomorphism between U_μ and an open and dense subset Θ_μ of R^d, with $d = n(p + s)$: $\Theta_\mu = \Phi_\mu(U_\mu)$

(4) $\bar{S}(n) \triangleq \bigcup_{i \le n} S(i) = \bar{U}_\mu$ if $|\mu| = n$.

Comment 1. Since U_μ, $|\mu| = n$, is open and dense in S(n), it follows that almost all points of S(n) are in U_μ for any such μ. The choice of a partition μ specifies a local coordinate system. Therefore, once n is chosen, a system in S(n) can be described in almost any coordinate system such that $|\mu| = n$. We shall see later that the partition μ and the corresponding coordinate vector τ_μ specify a pseudocanonical (SS, MFD, or ARMAX) form. The message therefore is that a given system of order n can be almost surely represented by any of the

$$\begin{bmatrix} n + p - 1 \\ p - 1 \end{bmatrix}$$

pseudocanonical forms corresponding to the

$$\begin{bmatrix} n + p - 1 \\ p - 1 \end{bmatrix}$$

partitions μ of n.

Comment 2. Let μ and ν be two partitions such that $|\mu| = |\nu| = n$. The intersection $U_\mu \cap U_\nu$ is also dense in S(n). A point in that intersection can be represented by either $\tau_\mu \in \Theta_\mu$ or $\tau_\nu \in \Theta_\nu$. Since S(n) is an analytic manifold, it follows that these two sets of coordinates are analytically related; that is the mapping

$$\phi_\mu \cdot \phi_\nu^{-1}: \phi_\nu(U_\mu \cap U_\nu) \to \phi_\mu(U_\mu \cap U_\nu) \tag{38}$$

is analytic.

Comment 3. S(n) can be covered by the

$$\begin{bmatrix} n + p - 1 \\ p - 1 \end{bmatrix}$$

open sets U_μ, $|\mu| = n$. Whether it can be covered by fewer sets is still an open question. It is known that if $p > 1$, no unique set U_μ can cover all of S(n). This has an important consequence: it means that the set of all systems in S(n) cannot be described by a unique identifiable representation.

Having described the structure of S(n), one can now think of the identification problem in the following terms. Estimate the order n, and then take any partition μ of n such that $|\mu| = n$ and compute the maximum likelihood estimate $\hat{\tau}_\mu$ of the corresponding vector τ_μ that completely specifies the system. However, this requires an algorithm that necessitates an I/O description of the point τ_μ. It turns out that the pseudocanonical SS, MFD, or ARMAX representations obtained from the τ_μ are not the simplest ones, as we shall see in the next sections. This is due to the overlap between the subsets U_μ for different μ. An alternative is to cover $S(n)$ by disjoint subsets V_μ, which can again be coordinatized by nonoverlapping coordinate systems. These will give rise to the somewhat simpler canonical SS, MFD, and ARMAX forms.

Definition 3. We call V_μ the subset of U_μ for which the rows (34) specified by μ are the first n linearly independent rows of $H_{1,\infty}[K]$.

Since the row r_{i,n_i+1} is now a linear combination of the basis rows above it, (35) is replaced by

$$r_{i,n_i+1} = \sum_{j=1}^{p} \sum_{l=1}^{n_{ij}} \alpha_{ijl} r_{jl}, \qquad i = 1, \ldots, p, \tag{39}$$

where

$$\begin{aligned} n_{ij} &\triangleq \min(n_i, n_j) && \text{if} \quad i \leq j, \\ &\triangleq \min(n_i + 1, n_j) && \text{if} \quad i > j. \end{aligned} \tag{40}$$

It follows, by the same argument as before, that any element $K(z) \in V_\mu$ can be mapped into a vector ρ_μ in $R^{d(\mu)}$ defined by the following coordinates:

$$\rho_\mu = \begin{cases} \alpha_{ijl}; & i, j = 1, \ldots, p; \quad l = 1, \ldots, n_{ij}, \\ k_l(i, j); & i = 1, \ldots, p; \quad j = 1, \ldots, s; \quad l = 1, \ldots, n_i, \end{cases} \tag{41}$$

where

$$d(\mu) = n(s + 1) + \sum_{i<j} \{\min(n_i, n_j) + \min(n_i, n_j + 1)\}. \tag{42}$$

We call Ψ_μ the mapping from K(z) into ρ_μ:

$$\Psi_\mu: V_\mu \to X_\mu \subset R^{d(\mu)}: K(z) \in V_\mu \to \rho_\mu = \Psi_\mu[K(z)] \in X_\mu. \quad (43)$$

The following result now holds about the V_μ (see, e.g., [4,14, 19]).

Result 3.

(1) The V_μ are disjoint, $V_\mu \subseteq U_\mu$ and $\bigcup_{|\mu|=n} V_\mu = S(n)$.

(2) Ψ_μ described in (43) is a homeomorphism between V_μ and an open and dense subset X_μ of $R^{d(\mu)}$, with $d(\mu)$ given by (42): $X_\mu = \Psi_\mu(V_\mu)$.

(3) X_μ is an open and dense subset of $R^{d(\mu)}$.

Comment 4. For $p > 1$, S(n) is partitioned into the

$$\begin{bmatrix} n + p - 1 \\ p - 1 \end{bmatrix}$$

disjoint sets V_μ, $|\mu| = n$, which are of different dimensions $d(\mu)$. Hence every system $\sigma \in S(n)$ belongs to one of the V_μ and therefore has a set of structure indices $\mu = (n_1, \ldots, n_p)$ attached to it. Those structure indices are usually called observability indices or left Kronecker indices; they determine the local coordinate system in which that system is described by the vector ρ_μ. These structure indices and these coordinates will in turn define canonical SS, MFD, or ARMAX forms, as we shall see in the next section. They form a complete system of independent invariants for K(z) (see [9,11,37]). Note that in all cases $d(\mu) \leq d = n(p + s)$. The canonical forms therefore will generally have fewer parameters than the corresponding pseudocanonical ones. This is one advantage of canonical forms.

Comment 5. If the first n rows of $H_{1,\infty}[K]$ are linearly independent, then μ has $n_1 = n_2 = \cdots = n_q = n_{q+1} + 1 = \cdots = n_p + 1$

for some q. Then $V_\mu = U_\mu$. Hence this subset is open and dense in S(n), and $\overline{V}_\mu = \overline{U}_\mu = \overline{S}(n)$. This particular V_μ is called the generic neighborhood, because generically a system will have Kronecker indices $n_1 = \cdots = n_q = n_{q+1} + 1 = \cdots = n_p + 1$ for some q. Hence, in practice, the generic neighborhood (and its corresponding canonical parametrization) is sufficient to represent almost any system. However, other nongeneric μ's (and their corresponding parametrizations) might be preferred for numerical reasons. For the generic μ, $d(\mu) = n(p + s)$; the other V_μ are mapped into spaces of lower dimension.

Recall now that a model structure was defined (see Section III, Definition 4) as a mapping from a parameter vector θ to a particular model M(θ) and that a structure was called globally identifiable if that mapping was injective for almost all θ in a subset D_M. Now we have just shown that (1) given a system $K(z) \in S(n)$, there exists a uniquely defined set of structure indices $\mu = (n_1, \ldots, n_p)$ and a uniquely defined mapping Ψ from K(z) to a parameter vector ρ_μ, and that (z) given a system $K(z) \in S(n)$, for almost any arbitrary set of structure indices $\mu = (n_1, \ldots, n_p)$ such that $\Sigma_1^p n_i = n$, there exists a uniquely defined mapping Φ_μ from K(z) to a parameter vector τ_μ. Therefore if we can now define model structures (in SS, MFD, or ARMAX form) that are entirely specified by the integer valued structure indices and the real-valued parameter vectors (ρ_μ or τ_μ), then these model structures will be identifiable since the sought-after inverse mappins are precisely Ψ_μ and Φ_μ. This is what we will set out to do in the next sections.

V. CANONICAL AND PSEUDOCANONICAL FORMS

We now describe SS, MFD, and ARMA canonical forms for a $p \times s$ transfer function $K(z)$ of order n, assuming that $K(z) \in V_\mu$, where $\mu = (n_1, \ldots, n_p)$. Recall that these Kronecker indices are determined by the linear dependence relations between the rows of $H_{1,\infty}[K]$. For simplicity of notation we assume that

$$y(t) = K(z)m(t) \tag{44}$$

rather than the original model (33) or (1). It is trivial, of course, to split $K(z)$ into $[G(z) \vdots H(z) - I]$, to replace $m(t)$ by

$$\begin{bmatrix} u(t) \\ e(t) \end{bmatrix},$$

and to add $e(t)$ to the right-hand side [see (33)]. An ARMA model would then be converted back into an ARMAX model. However, at this stage the only issue is one of parametrizing a SS, MFD, or ARMA model of a strictly causal rational transfer function.

A. *CANONICAL STATE-SPACE FORM*

A canonical state-space model A, B, C such that $K(z) = C(zI - A)^{-1}B$ is obtained from the complete set of invariants defining the vector ρ_μ [see (41)] as follows:

$$\underset{(p\times n)}{C} = \left[\begin{array}{cccc|cccc|c|cccc} 1 & 0 & \cdots & 0 & 0 & & \cdots & 0 & & 0 & & \cdots & 0 \\ 0 & & & & 1 & 0 & \cdots & 0 & & & & & \\ & & & & 0 & & & & & \vdots & & & \vdots \\ \vdots & & & \vdots & \vdots & & & \vdots & \cdots & & & & \\ & & & & & & & & & 0 & & & 0 \\ 0 & & \cdots & 0 & 0 & & \cdots & 0 & & 1 & 0 & \cdots & 0 \end{array}\right], \tag{45a}$$
$$\qquad\quad \underbrace{\qquad\qquad}_{n_1} \qquad \underbrace{\qquad\qquad}_{n_2} \qquad\qquad \underbrace{\qquad\qquad}_{n_p}$$

$$\underset{(n\times n)}{A} = \left[\begin{array}{ccc:ccc:c:ccc}
0 & & & 0 & \cdots & 0 & & 0 & \cdots & 0 \\
\vdots & I_{n_1-1} & & \vdots & & \vdots & \cdots & & & \vdots \\
0 & & & 0 & \cdots & 0 & & 0 & & 0 \\
\alpha_{111} & \cdots & \alpha_{11n_1} & \alpha_{121} \cdots \alpha_{12n_{12}} & & 0 \cdots 0 & & \alpha_{1p1} \cdots \alpha_{1pn_{1p}} & & 0 \cdots 0 \\
\hdashline
 & \vdots & & & \vdots & & \cdots & & \vdots & \\
\hdashline
0 & \cdots & 0 & 0 & \cdots & 0 & & 0 & & \\
\vdots & & \vdots & \vdots & & \vdots & & \vdots & I_{n_p-1} & \\
0 & \cdots & 0 & 0 & \cdots & 0 & & 0 & & \\
\alpha_{p11} \cdots \alpha_{p1n_{p1}} & & 0 \cdots 0 & \alpha_{p21} \cdots \alpha_{p2n_{p2}} & & 0 \cdots 0 & & \alpha_{pp1} & \cdots & \alpha_{ppn_p}
\end{array}\right], \tag{45b}$$

$$\underset{(n\times s)}{B} = \left[\begin{array}{ccc}
k_1(1, 1) & \cdots & k_1(1, s) \\
\vdots & & \vdots \\
k_{n_1}(1, 1) & \cdots & k_{n_1}(1, s) \\
\hdashline
 & \vdots & \\
\hdashline
k_1(p, 1) & \cdots & k_1(p, s) \\
\vdots & & \vdots \\
k_{n_p}(p, 1) & \cdots & k_{n_p}(p, s)
\end{array}\right]. \tag{45c}$$

Note that A and C have a very specific structure while B is fully parametrized. In an identification context, once the structure indices n_i have been estimated, the structure of A and C is completely specified by the 0 and 1 elements, while the α_{ijk} and $k_l(i, j)$ are free parameters. Any arbitrary state-space representation of an nth order system with Kronecker indices $(n_1, \ldots, n_p)$ can be transformed to this canonical form by a similarity transformation (see, e.g., [11] or [38] for details). Finally, note that this form has $d(\mu)$ parameters.

B. CANONICAL MFD FORM

A canonical MFD form P(z), Q(z) such that $K(z) = P^{-1}(z)Q(z)$ is obtained as follows from ρ_μ.

Let

$$\underset{p\times p}{P(z)} \triangleq [p_{ij}(z)] \quad \text{and} \quad \underset{p\times s}{Q(z)} \triangleq [q_{ij}(z)].$$

Then

$$p_{ii}(z) = z^{n_i} - \alpha_{iin_i} z^{n_i - 1} - \cdots - \alpha_{ii1}, \tag{46a}$$

$$p_{ij}(z) = -\alpha_{ijn_{ij}} z^{n_{ij}-1} - \cdots - \alpha_{ij1} \quad \text{for} \quad i \neq j, \tag{46b}$$

$$q_{ij}(z) = \beta_{ijn_i} z^{n_i - 1} + \cdots + \beta_{ij1}, \tag{46c}$$

where the coefficients β_{ijl} are bilinear functions of the coefficients α_{ijl} and $k_l(i, j)$ obtained as follows. Let

$$\underset{(n\times s)}{G} \triangleq \begin{bmatrix} G_1 \\ \vdots \\ G_p \end{bmatrix}, \qquad \underset{(n_i\times s)}{G_i} \triangleq \begin{bmatrix} \beta_{i11} & & \beta_{is1} \\ \vdots & & \vdots \\ \beta_{i1n_i} & \cdots & \beta_{isn_i} \end{bmatrix},$$

$$\underset{(n\times n)}{M} \triangleq [M_{ij}], \quad i, j = 1, \ldots, p, \tag{47}$$

with

$$\underset{(n_i\times n_i)}{M_{ii}} \triangleq \begin{bmatrix} -\alpha_{ii2} & -\alpha_{ii3} & \cdots & -\alpha_{iin_i} & 1 \\ -\alpha_{ii3} & & \cdots & 1 & 0 \\ \vdots & & & & \vdots \\ -\alpha_{iin_i} & 1 & \cdot^{\cdot^{\cdot}} & & \\ 1 & 0 & \cdots & & 0 \end{bmatrix},$$

$$\underset{(n_i \times n_j)}{M_{ij}} \triangleq \begin{bmatrix} -\alpha_{ij2} & -\alpha_{ij3} & \cdots & -\alpha_{ijn_{ij}} & 0 & \cdots & 0 \\ -\alpha_{ij3} & \cdots & & 0 & & \cdots & 0 \\ \vdots & & & & & & \\ -\alpha_{ijn_{ij}} & & \cdot^{\cdot^{\cdot}} & & & & \vdots \\ 0 & & & & & & \\ \vdots & & & & & & \\ 0 & & & \cdots & & & 0 \end{bmatrix}. \tag{48}$$

Then

$$G = MB. \tag{49}$$

See [11] for a proof. This canonical form also has $d(\mu)$ parameters. It has the following properties, which actually define its structure:

(i) The polynomials on the main diagonal of $P(z)$ are monic with

$$\deg(p_{ii}) = n_i, \tag{50a}$$

(ii) $$\deg(p_{ij}) \leq \deg(p_{ii}) \quad \text{for} \quad j \leq i, \tag{50b}$$

$$\deg(p_{ij}) < \deg(p_{ii}) \quad \text{for} \quad j > i, \tag{50c}$$

$$\deg(p_{ji}) < \deg(p_{ii}) \quad \text{for} \quad j \neq i, \tag{50d}$$

(iii) $\deg(q_{ij}) < \deg(p_{ii})$ and $P(z)$, $Q(z)$ are left coprime. (50e)

The form $P(z)$, $Q(z)$ with the properties (50) was first proposed by Guidorzi [10] and is called the canonical echelon form in econometrics (see [25]). In the control engineering literature a closely related canonical MFD is called the canonical echelon form (see [29]). It is obtained from the Guidorzi form by permuting the rows of $P(z)$ [and correspondingly of $Q(z)$] such that in the transformed $\overline{P}(z)$ (1) the row degrees are arranged in increasing order and (2) if in $P(z)$ $n_i = n_j$ with $i < j$, then in $\overline{P}(z)$ the ith row of $P(z)$ is above the jth row of $P(z)$.

If we denote by P_{hc} the highest column degree coefficient matrix of P(z) [i.e., the matrix whose columns are the coefficients of the highest power of z in each column of P(z)] and by P_{hr} the highest row degree coefficient matrix, then it follows easily from the properties (50) that

(i) $P_{hc} = I_p$, (51a)

(ii) P_{hr} is lower triangular with unit diagonal elements. (51b)

It follows that P(z) is both column reduced and row reduced (see [29]) with row degrees and column degrees equal to $n_1, \ldots, n_p$. It also follows that

$$\deg \det P(z) = \sum_{1}^{p} n_i = n = \text{order of } K(z). \tag{52}$$

In an identification context, once the Kronecker indices n_i have been estimated, the structure of P(z) and Q(z) is completely specified by (46) or, equivalently, by the degree relations (50), where the α_{ijk} and β_{ijk} are free parameters to be estimated from the data; note that the number of free parameters β_{ijk} is identical to the number of parameters $k_l(i, j)$ in ρ_μ, that is, ns.

C. *CANONICAL ARMA FORM*

Using (25) it is easy to obtain a canonical ARMA model from the echelon MFD model:

$$[\overline{P}(D) \vdots \overline{Q}(d)] = M(D)[P(z) \vdots Q(z)], \tag{53}$$

where $M(D) \triangleq \text{diag}\{D^{n_1}, \ldots, D^{n_p}\}$. It is easy to see, using (50), that $\overline{P}(D)$, $\overline{Q}(D)$ have the following properties:

(i) $\overline{P}(D)$ and $\overline{Q}(D)$ are left coprime. (54a)

(ii) $\bar{P}(0) = P_{hr}$ is lower triangular and nonsingular. (54b)

(iii) The row degrees of $[\bar{P}(D) \vdots \bar{Q}(D)]$ are $n_1, \ldots, n_p$. (54c)

The parameters in this canonical form are identical to those appearing in the echelon MFD form, and their positions are again determined by the Kronecker indices. It is therefore identifiable. Notice that the row degrees of $\bar{P}(D)$ are not necessarily $n_1, \ldots, n_p$ (this depends on the particular system, i.e., on the values of the coefficients α_{ijk}), while the column degrees of $\bar{P}(D)$ are generically equal to $r \triangleq \max_i\{n_i\}$. Recall also Section II, Comment 2. One major disadvantage of this ARMA canonical form is that $\bar{P}(0) \neq I$; that is, we cannot write y(t) in the form (24). One way to obtain an ARMA (or ARMAX) form such as (24) is to multiply $[\bar{P}(D) \vdots \bar{Q}(D)]$ to the left by P_{hr}^{-1}. However, this increases some of the row degrees of $\bar{P}(D)$ and $\bar{Q}(D)$ (and hence the lag structure of the model) and therefore increases the number of parameters in the model. In fact, it can be shown [31] that a system in S(n) with Kronecker indices $n_1, \ldots, n_p$ can in general not be represented by an ARMA model

$$\bar{P}(D)y(t) = \bar{Q}(D)m(t), \quad \bar{P}(0) = I, \tag{55}$$

such that the row degrees of $[\bar{P}(D) \vdots \bar{Q}(D)]$ are $n_1, \ldots, n_p$ and that an ARMA model of the form (55) will have more than $d(\mu)$ parameters. Moreover, ARMA models of the form (55) will generically represent systems whose order is a multiple of p, the dimension of y(t) (see [31] for details).

We return now to the mapping (43). If we denote by $X_\mu(SS)$ [resp. $X_\mu(MFD)$, $X_\mu(ARMA)$] the set of all free parameters in the canonical SS (resp. MFD, ARMA) form, then it follows from

Section IV, Result 3 and the fact that the parameters of these three canonical forms are bijectively related to the components of $\rho_\mu \in X_\mu$, that the sets X_μ, $X_\mu(SS)$, $X_\mu(MFD)$, and $X_\mu(ARMA)$ are homeomorphic. Finally, note that the canonical forms we have described here are just one possible set of canonical forms. Using different (but uniquely defined) selection rules for the basis vectors of $H_{1,\infty}[K]$, one can obtain a number of other SS, MFD, and ARMA forms (see, e.g., [7,29,38,39]).

The main disadvantage of canonical forms for identification is that one has to estimate the Kronecker indices. In Section VII we shall briefly describe different methods for doing this, but in any case it is a time-consuming and numerically sensitive procedure. An alternative is to use pseudocanonical forms; this requires the estimation of only one integer-valued parameter, the order n. For almost every system $\sigma \in S(n)$, any set of structure indices $\mu = (n_1, \ldots, n_p)$ such that $\Sigma_1^p h_i = n$ can then be used to define the structure of a pseudocanonical form. We now describe SS, MFD, and ARMA pseudocanonical forms, which are very similar to the canonical forms just described.

D. *PSEUDOCANONICAL STATE-SPACE FORM*

Let $K(z) \in U_\mu$. Then a pseudocanonical SS form for K(z) is obtained from the parameter vector τ_μ by taking C as in (45a), B as in (45c), and

$$\underset{(n\times n)}{A} \triangleq [A_{ij}], \quad \text{with} \quad \dim A_{ij} = n_i \times n_j, \tag{56}$$

where the $[A_{ii}]$ are as in (45b) and where

$$\underset{(n_i\times n_j)}{A_{ij}} = \begin{bmatrix} 0 & \cdots & 0 \\ \vdots & & \vdots \\ 0 & \cdots & 0 \\ \alpha_{ij1} & \cdots & \alpha_{ijn_j} \end{bmatrix}. \tag{57}$$

(See, e.g., [23] for a derivation of this form.) Note that the free parameters are exactly the coordinates of τ_μ, which uniquely describe the system in the coordinate space defined by μ. Hence this form is identifiable.

E. PSEUDOCANONICAL MFD FORM

A pseudocanonical MFD form for $P(z)$, $Q(z)$ is obtained from the coordinates of τ_μ. [Compare with (46)-(50).]

$$p_{ii}(z) = z^{n_i} - \alpha_{iin_i} z^{n_i - 1} - \cdots - \alpha_{ii1}, \tag{58a}$$

$$p_{ij}(z) = -\alpha_{ijn_j} z^{n_j - 1} - \cdots - \alpha_{ij1} \quad \text{for} \quad i \neq j, \tag{58b}$$

$$q_{ij}(z) = \beta_{ij\rho_i} z^{\rho_i - 1} + \cdots + \beta_{ij1}, \tag{58c}$$

where $\rho_i \triangleq$ ith row degree of $P(z) = \max(n_i, \max_j\{n_j\} - 1)$. The bilinear relations (49) between the β_{ijl} and $k_l(i, j)$ still hold, with B as before but with

$$\underset{(\rho_i \times s)}{G_i} \triangleq \begin{bmatrix} \beta_{i11} & & \beta_{is1} \\ \vdots & & \vdots \\ \beta_{i1\rho_i} & \cdots & \beta_{is\rho_i} \end{bmatrix}, \tag{59}$$

$$\underset{(\rho_i \times n_i)}{M_{ii}} \triangleq \begin{array}{l} \\ \\ \\ \\ \\ \rho_i - n_i \left\{ \begin{array}{l} \\ \\ \\ \end{array} \right. \end{array} \begin{bmatrix} -\alpha_{ii2} & \cdots & \alpha_{iin_i} & 1 \\ \vdots & & & 0 \\ -\alpha_{iin_i} & \cdot^{\cdot^{\cdot}} & & \\ & & & \vdots \\ 1 & & & \\ 0 & & & \\ \vdots & & & \\ 0 & \cdots\cdots & & 0 \end{bmatrix},$$

$$\underset{(\rho_i \times n_j)}{M_{ij}} \triangleq \begin{bmatrix} -\alpha_{ij2} & \cdots & -\alpha_{ijn_j} & 0 \\ \vdots & & \cdot^{\cdot^{\cdot}} & \\ -\alpha_{ijn_j} & \cdot^{\cdot^{\cdot}} & & \vdots \\ \cdot & & & \\ 0 & & & \\ \vdots & & & \\ 0 & & \cdots & 0 \end{bmatrix}. \quad \rho_i - n_j + 1 \text{ (last rows)} \tag{60}$$

The relationship (49) for this pseudocanonical form, with G and M as just described, was derived independently in [21,40,41]. The pseudocanonical form has the following properties:

(i) The p_{ii} are monic with $\deg(p_{ii}) = n_i$. (61a)

(ii) $\deg(p_{ji}) < \deg(p_{ii}) = n_i$ for $j \neq i$. (61b)

(iii) $P(z)$ and $Q(z)$ are left coprime, and $\deg \det P(z) = n$. (61c)

It follows from (61b) that $P(z)$ is column reduced with column degrees $(n_1, \ldots, n_p)$. However, it is not row reduced and this could make some of the I/O relations apparently nonstrictly causal if $\rho_i - 1 \geq n_i$. This will be the case if $\max_i\{n_i\} - \min_i\{n_i\} \geq 2$. The problem arises because the parameters β_{ijl} are not all free; the relationship (49) is not a bijection between the $k_l(i, j)$ and the β_{ijl} here (as it was in the canonical echelon form), since the number of elements in G is larger than that in B. It was shown in [19] that the following set of $n(p + s)$ parameters may be chosen as free:

$$\begin{cases} \alpha_{ijl}; & i, j = 1, \ldots, p; \quad l = 1, \ldots, n_j, \\ \beta_{ijl}; & i = 1, \ldots, p; \quad j = 1, \ldots, s; \quad l = 1, \ldots, n_i, \end{cases} \tag{62}$$

Compare with (36). The other β_{ijl} are then nonlinear combinations of the parameters in (62). When these nonlinear constraints are taken into account, the apparent nonstrict causalities disappear. In an identification context, if the α_{ijl} and β_{ijl} are estimated independently, the constraints will not be exactly satisfied, and noncausal relations may appear. However, Correa and Glover have explicitly computed these constraints and they have shown that by reordering the output variables such that $n_1 \geq \cdots \geq n_p$ (this is always possible), one can treat the I/O relations one by one and successively eliminate the dependent β_{ijl} [24].

F. PSEUDOCANONICAL ARMA FORM

Because $P(z)$ is not row proper, P_{hr} can be singular. Therefore, if we multiply $[P(z) \vdots Q(z)]$ to the left by $M(D)$ as in (53), there is no guarantee that $\overline{P}(0)$ is nonsingular, making the ARMA pseudocanonical form difficult to use. In Section VI we shall present an alternative ARMA form with $\overline{P}(0) = I$. While this form has more than $n(p + s)$ parameters [recall that $n(p + s)$ is the dimension of the space of the overlapping subsets U_μ] and requires a larger number of integer-valued structure indices for its definition, it is identifiable. Other identifiable ARMA forms that have more than the minimum number of parameters have been proposed in [23].

The choice between using either the canonical forms or the pseudocanonical forms described in this section is a fairly subjective one. Canonical forms may have slightly fewer parameters leading to more efficient estimates; on the other hand, the structure estimation step of the identification requires the estimation of p structure indices while, with pseudocanonical forms, only the order must be estimated. If pseudocanonical forms are used and if the (arbitrarily chosen) set of structure indices leads to a numerically ill-conditioned parametrization, a coordinate transformation can be used [see (38)] to move to a better-conditioned parametrization [16]. If pseudocanonical forms are used, it is always a good idea to start with the generic μ; see Comment 5.

VI. OTHER IDENTIFIABLE PARAMETRIZATIONS

The canonical and pseudocanonical parametrizations of Section V were directly derived from the coordinates τ_μ (resp. ρ_μ) of K(z) in the coordinate spaces spanning U_μ (resp. V_μ). The number of free parameters in these parametrizations is entirely determined by the order of the system (resp. the Kronecker indices) and that number is minimal; it equals the dimension of the space U_μ (resp. V_μ). However, these parametrizations are by no means the only identifiable ones. In this section we describe some other identifiable parametrizations; they will most often have more free parameters than the ones described earlier, but they have some other useful properties.

A. *FULLY PARAMETRIZED ARMA MODELS*

We consider the class of stable rational strictly proper K(z) of dimension $p \times s$ that can be modeled as

$$\overline{K}(D) \triangleq K(D^{-1}) = \overline{P}^{-1}(D)\overline{Q}(D), \tag{63}$$

where

$$\text{(i)} \quad \overline{P}(D) = I_p + \overline{P}_1 D + \cdots + \overline{P}_u D^u, \tag{64a}$$

$$\overline{Q}(D) = \overline{Q}_1 D + \cdots + \overline{Q}_v D^v, \tag{64b}$$

$$\text{(ii)} \quad \overline{P}(D) \text{ and } \overline{Q}(D) \text{ are left coprime,} \tag{64c}$$

$$\text{(iii)} \quad \text{rank}[\overline{P}_u \vdots \overline{Q}_v] = p. \tag{64d}$$

We call $S(u, v)$ the set of all $K(z)$ that can be modeled by (63)-(64) with prescribed degrees u, v, and we denote by $\Theta(u, v)$ the set of all parameters in $\bar{P}(D)$ and $\bar{Q}(D)$ that are not identically 0 or 1 and for which (64) holds. The set $S(u, v)$ has the following properties (see [20,25,26,42,43]).

Result 1.

(1) $S(u, v)$ is mapped homeomorphically into an open set $\Theta(u, v) \subset R^d$, where $d = p(p \times u + s \times v)$ by the mapping

$$\phi_{u,v}: S(u, v) \to \Theta(u, v): K(z) \in S(u, v) \to \tau_{u,v}[K(z)] \in \Theta(u, v), \tag{65}$$

where $\tau_{u,v}$ is the vector of the coefficients appearing in (64a) and (64b), and hence $\Theta(u, v)$ is identifiable.

(2) $\{S(u, v),\ u, v \in Z_+\}$ is not a cover of $S(n)$: that is, there exist $K(z) \in S(n)$ for which no u, v exist such that $K(z) \in S(u, v)$.

(3) The $S(u, v)$ are not disjoint: that is, a given $K(z)$ can be in $S(u_1, v_1)$ and $S(u_2, v_2)$ for $(u_1, v_1) \neq (u_2, v_2)$.

Note that d here will always be larger than the dimension $n(p + s)$ of the overlapping submanifolds of Section IV and V. In addition, the order of such models will generically be a multiple of p [31]. The problem raised in the preceding paragraph (2) can be eliminated if, instead of prescribing the highest column degrees (u, v) of $\bar{P}(D)$ and $\bar{Q}(D)$, we prescribe the column degrees $(u_1, \ldots, u_p; v_1, \ldots, v_s)$ of each column of $[\bar{P}(D) \vdots \bar{Q}(D)]$. We denote by a_i (resp. b_i) the vector of coefficients of D^{u_i} (resp. D^{v_i}) in the ith column of $\bar{P}(D)$ [resp. $\bar{Q}(D)$]. Now we denote by $S(u_1, \ldots, u_p; v_1, \ldots, v_s)$ the set of all $K(z)$ that can be modeled by (63), where $\bar{P}(D)$ and $\bar{Q}(D)$ obey

(64a)-(64c) with $u = \max_i u_i$, $v = \max_i v_i$, and where

$$\text{rank}[a_1, \ldots, a_p; b_1, \ldots, b_s] = p \tag{66}$$

for prescribed column degrees $(u_1, \ldots, u_p; v_1, \ldots, v_s)$, and we denote by $\Theta(u_1, \ldots, u_p; v_1, \ldots, v_s)$ the set of all parameters in these $\overline{P}(D)$, $\overline{Q}(D)$ that are not identically 0 and 1. We then have the following result.

Result 2 [20,26,44]

(1) $S(u_1, \ldots, u_p; v_1, \ldots, v_s)$ is mapped homeomorphically onto an open and dense subset $\Theta(u_1, \ldots, u_p; v_1, \ldots, v_s)$ of R^d, where $d = p\left(\Sigma_1^p u_i + \Sigma_1^s v_i\right)$, and hence $\Theta(u_1, \ldots, u_p; v_1, \ldots, v_s)$ is identifiable.

(2) For every $K(z) \in S(n)$, there exists $(u_1, \ldots, u_p; v_1, \ldots, v_s)$ such that $K(z) \in S(u_1, \ldots, u_p; v_1, \ldots, v_s)$.

(3) The $S(u_1, \ldots, u_p; v_1, \ldots, v_s)$ are not disjoint: that is, a $K(z)$ can be modeled uniquely by ARMA models having different u_i, v_i, each obeying the constraints (64a), (64c), and (66).

The most detailed discussion of this identifiable model structure is given in [20]. A major disadvantage of this form is that p + s integer-valued parameters must be prescribed.

B. A "SCALAR" ARMA MODEL

A commonly used representation for K(z) is

$$\overline{p}(D)y(t) = \overline{Q}(D)m(t), \tag{67}$$

where

(i) $\overline{p}(D) = 1 + \overline{p}_1 D + \cdots + \overline{p}_u D^u$ is a scalar polynomial with $\overline{p}_u \neq 0$. (68a)

(ii) $\bar{Q}(D) = \bar{Q}_1 D + \cdots + \bar{Q}_v D^v$ has dimension $p \times s$ with

$$\bar{Q}_v \neq 0. \tag{68b}$$

(iii) $\bar{p}(D)I$ and $\bar{Q}(D)$ are left coprime. (68c)

We call $S_{sc}(u, v)$ the set of all stable rational strictly proper $K(z)$ that can be modeled by (67) under the constraints (68). We then have the following result.

Result 3 [42].

(1) $S_{sc}(u, v)$ is mapped homeomorphically onto an open and dense subset $\Theta_{sc}(u, v)$ of R^d with $d = u + p \times s \times v$, and hence $\Theta_{sc}(u, v)$ is identifiable.

(2) $S_{sc}(u, v)$ covers $S(n)$.

This last result follows immediately from the fact that the form (67) is obtained by taking $P(z) = z^r \bar{p}(D)$, with $r \triangleq \max(u, v)$, as the least common denominator of the elements of $K(z)$. Notice finally that the form (67) contains in general more parameters than the canonical or pseudocanonical forms.

C. *ELEMENTARY SUBSYSTEM REPRESENTATIONS*

The elementary subsystem (ESS) representation for multivariable linear systems, introduced in [45]-[46], is based on a decomposition of the monic least common denominator $p(z)$ of the elements of $K(z)$ into irreducible first- and second-degree polynomials:

$$p(z) = \prod_{i=1}^{n_r} p_{ri}(z) \prod_{j=1}^{n_c} p_{cj}(z), \tag{69}$$

where

$$p_{ri}(z) = z + a_i, \quad p_{cj}(z) = z^2 + b_{1j}z + b_{2j}. \tag{70}$$

It is assumed that the poles of K(z) have multiplicity one. Collecting all terms associated with $P_{ri}(z)$ into a matrix $K_{ri}(z)$ and all terms associated with $p_{cj}(z)$ into a matrix $K_{cj}(z)$, one can then write K(z) as the sum of these partial fraction matrices (PFM):

$$K(z) = \sum_{i=1}^{n_r} K_{ri}(z) + \sum_{j=1}^{n_c} K_{cj}(z). \tag{71}$$

The PFM $K_{ri}(z)$ [resp. $K_{cj}(z)$] can then be realized as a direct sum of first-order (resp. second-order) elementary subsystems using dyadic decompositions of the numerator matrices. Now K(z) is finally realized as a direct sum of these direct sums. The procedure is explained in great detail in [46], where a structure estimation scheme is also proposed. The authors claim that this elementary subsystem structure is identifiable; however, it is not clear that this will be the case if several elements of K(z) have the same pole, unless some additional rules are imposed to make the dyadic decompositions unique. On the other hand, the advantage of this structure is that it is often close to the physical model of the system or its subsystems. Since the poles of K(z) are directly estimated, stability can easily be checked, or stability constraints can be introduced.

VII. THE ESTIMATION OF THE STRUCTURE

In this section we shall briefly survey some important theoretical results on structure estimation and point to a number of practical methods that have been proposed for the estimation of the order n or the Kronecker indices $(n_1, \ldots, n_p)$ or the lag lengths (u, v). In the case where pseudocanonical forms are used, any set of structure indices $\mu = (n_1, \ldots, n_p)$ adding up to n can in principle be used; however, also in this case methods have been suggested for selecting a partition μ that leads to a numerically better-conditioned estimation algorithm. There is a very abundant literature on structure estimation, mainly originating from statisticians. For reasons of space, we shall be able to present only the main thrust of the results without going into technical details. Our starting point will be the model (1) again. We shall sometimes specialize to methods that do not allow deterministic inputs or others that consider only deterministic inputs. For brevity of notation, we shall also sometimes use $K(z) \triangleq [G(z) \vdots H(z) - I]$ as before.

A. *RESULTS USING THE MAXIMUM LIKELIHOOD*

We first discuss some important consistency results for maximum likelihood estimation of parameters. We will assume that the model (1) is subjected to the conditions (2) and that the u(t) are observable. We shall denote by F_t the σ-algebra of events determined by $\{y(s), s \leq t\}$; equivalently, since u(t) is observable and (2e) is assumed, F_t is determined by $\{e(s), s \leq t\}$. It is now assumed that there exists a true system $K_0(z) \triangleq [G_0(z) \vdots H_0(z) - I]$ with innovations e(t) obeying the following assumptions:

$$\text{(i)} \quad E\{e(t)|F_{t-1}\} = 0 \text{ a.s.}, \qquad E\left\{e(t)e^T(t)|F_{-\infty}\right\} = \Sigma_0. \tag{72a}$$

$$\text{(ii)} \quad e(t) \text{ is ergodic.} \tag{72b}$$

$$\text{(iii)} \quad E\left\{[e_j(t)]^4\right\} < \infty \quad \text{for } j = 1, \ldots, p. \tag{72c}$$

Conditions (72a) ensure that $e(t)e^T(t)$ is purely nondeterministic and that the best predictor of $y(t)$ given F_{t-1} is the best linear predictor. We now denote by U any of the sets of transfer functions $K(z)$ described in Sections IV through VI by $\bar{U}$ its closure w.r.t. T_{pt}, by Θ the parameter set of a corresponding identifiable model structure, by Φ the mapping from U to Θ such that $\Phi(K(z)) = \tau \in \Theta$, and by Π the inverse mapping such that $\Pi(\tau) = K(z) \in U$ for $\tau \in \Theta$. For example, if U is taken as U_μ, then $K(z)$ is the set of transfer functions such that $K(z) \in U_\mu$, $\bar{U} = \bar{S}(n)$, where $n = |\mu|$ by Result 2 in Section IV (4), Θ could be taken as $\Theta_\mu(SS)$ (see Section V), any element $K(z) \in U_\mu$ is then mapped into $\tau_\mu = \Phi_\mu(K(z))$ [see (37)] and $\Pi(\tau_\mu) = K(z)$. Setting the initial values of $u(\cdot)$ and $y(\cdot)$ to zero, and denoting $V_N = [v^T(1), \ldots, v^T(N)]^T$, where $v(t) \triangleq y(t) - \Sigma_1^t G_i u(t - i)$ and

$$\Gamma_N(\tau, \Sigma) \triangleq E\left\{V_N V_N^T\right\}, \tag{73}$$

then the likelihood function is given by

$$L_N(\tau, \Sigma) = \frac{1}{N} \log \det \Gamma_N(\tau, \Sigma) + \frac{1}{N} V_N^T \Gamma_N^{-1}(\tau, \Sigma) V_N. \tag{74}$$

Now the important point is that $L_N(\tau, \Sigma)$ depends on the parameter vector τ only through $\Pi(\tau) = K(z)$. Therefore the likelihood function $L_N(\tau, \Sigma)$ can be considered as "coordinate free;" the particular parametrization is unimportant. We then have the following important consistency result [3].

Result 1. Assume that $y(t)$ is generated by an ARMAX process (10) with the assumptions (2) and (72) and assume that $K_0(z) \in \bar{U}$. If $\hat{\tau}_N$, $\hat{\Sigma}_N$ are the MLEs obtained by optimizing $L_N(\tau, \Sigma)$ over

$\bar{U} \times \{\Sigma | \Sigma > 0\}$, and if $\hat{K}_N(z) = \Pi(\hat{\tau}_N)$, then

$$\hat{K}_N(z) \to K_0(z) \quad \text{in } T_{pt} \text{ a.s.,} \quad \text{and} \quad \hat{\Sigma}_N \to \Sigma_0 \quad \text{a.s.} \tag{75}$$

This result has the following consequences.

(1) If, say, U is taken as U_μ, $|\mu| = n$, and if $K_0(z) \in S(j)$, $j \leq n$, then $\bar{U} = \bar{S}(n) = \bigcup_{j\leq n} S(j)$, so that the MLE is consistent even when the true order is exceeded.

(2) Result 1 is about consistency of transfer function estimates, not about consistency of the parameter vector $\hat{\tau}_N$. If U is taken as U_μ and $K_0(z) \in U_\mu$, then $\hat{\tau}_N \to \tau_0$ a.s., but other situations can arise when $K_0(z) \notin U_\mu$ (see [25]).

Result 1 makes the crucial assumption that $K_0(z) \in \bar{U}$. This almost amounts to saying that the structure of $K_0(z)$ is known. We now turn to results on structure estimation. A maximum likelihood criterion cannot be used for, say, order estimation for the following reason. If n_0 is the true order and if $n_1 > n_0$, then $\bar{S}(n_0) \subset \bar{S}(n_1)$ and $S(n_1)$ is dense in $\bar{S}(n_1)$; therefore the MLE over $\bar{S}(n_1)$ will almost surely be attained in $S(n_1)$; MLE will almost always overestimate the order.

Order estimation criteria therefore add a penalty term on the dimension d of the parameter space. They are generally obtained by minimizing a criterion of the form

$$A_N(n) = \log \det \hat{\Sigma}_N(n) + d(n)[C(N)/N], \qquad n = 0, 1, \ldots, n_{max}, \tag{76}$$

where $\hat{\Sigma}_N(n)$ is the MLE of Σ_0 over $\bar{S}(n) \times \{\Sigma | \Sigma > 0\}$, n_{max} is the maximum order to be considered, $d(n)$ is the dimension of the parameter space, N is the number of observations, and $C(N)$ can take different values. Such criteria were first proposed by

Akaike. If $C(N) = 2$, $A_N(n)$ is called AIC [47]; if $C(N) = C \log N$, $A_N(n)$ is called BIC [48]. Other criteria, based on minimum description length of data, are due to Rissanen [49,50]; a third term is added to the expression (76).

The criterion (76) has been expressed as a function of the order n. It can just as well be expressed as a function of the set of Kronecker indices $\mu = (n_1, \ldots, n_p)$ when the search is performed over the disjoint canonical forms instead of the overlapping pseudocanonical forms:

$$A_N(\mu) = \log \det \hat{\Sigma}_N(\mu) + d(\mu)[C(N)/N], \qquad |\mu| \leq n_{max}. \qquad (77)$$

Under a reasonable set of assumptions, the following results have been obtained in [26] through [28].

Result 2.

(1) For $U = V_\mu$ (i.e., disjoint neighborhoods), BIC gives strongly consistent estimates $\hat{\mu}_N$ of the Kronecker indices.

(2) For $U = U_\mu$ (i.e., overlapping neighborhoods), BIC gives a strongly consistent estimate $\hat{n}_N$ of the true order n_0.

(3) AIC is not consistent in cases (1) and (2) (it overestimates the order).

However, in practice there will be no true order or Kronecker indices: that is, the true system will not have a rational transfer function. AIC seems to be directed at this situation. Using a particular criterion of optimality for this situation, Shibata has shown that AIC has some optimal properties for spectral estimation using autoregressive models [51,52].

For the practical applicability of these structure estimation results, it is important to observe that the criteria (76)-(77) use the MLE $\hat{\Sigma}_N(n)$ or $\hat{\Sigma}_N(\mu)$. Since the search for the Kronecker indices is to be performed over a large number of candidate models, this represents a formidable computational task. One way of reducing the effort is to search first for the largest Kronecker index by performing the search over $\mu_i = (n_1, \ldots, n_p)$, where $n_1 = \cdots = n_p = i$ for increasing i. If $\hat{\mu}_l = \min_{i=1,2,\ldots} A_N(\mu_i)$, then l is an estimate of the largest Kronecker index. Subsequently a much smaller number of V_μ need be examined to estimate the remaining Kronecker indices. However, this method still requires the minimization of a number of likelihood functions. Therefore other simpler methods have been sought, which we will briefly describe next.

B. OTHER METHODS FOR THE ESTIMATION OF THE KRONECKER INDICES

Hannan and Kavalieris [26] (see also [53]) have proposed a method that minimizes the criterion (77), but where the MLE $\hat{\Sigma}_N(\mu)$ is computed cheaply using only linear equations. The method assumes that K(z) belongs to a generic neighborhood, that is, that $n_1 = n_2 = \cdots n_q = n_{q+1} + 1 = \cdots = n_p + 1$ for some q. This is not a severe limitation, since generic neighborhoods are dense in $\bar{S}(n)$ (see Comment 5). On the other hand, it greatly reduces the number of candidate models. It is also assumed that y(t) is generated by an ARMAX model. The method uses a three-stage procedure inspired by Durbin [54]. In stage I an autoregressive model, whose order increases with the number of data, is fitted to the data; it is used to compute estimates of the innovations. Using these estimates, canonical and generic ARMAX models are fitted in stage II; the residual variances from these models are used in (77) to estimate μ. Once the model structure is chosen, stage III computes MLEs of the parameters. It is shown that this procedure gives consistent estimates of the Kronecker indices under assumptions that are only slightly stronger than those required for Result 2.

A number of other simple methods have been proposed for the estimation of the Kronecker indices; however, no consistency results are available for these methods. They are all based on the fact that the Kronecker indices can be inferred from the linear dependence relations on the rows of $H_{1,\infty}[K]$. The problem is that K(z) is not known. One procedure is to first estimate the K_i by a long autoregression. If there are no deterministic inputs, then the covariance matrix between the vectors of "future" outputs and "past" outputs can be used to establish the linear dependence relationships; the rows of that block-Hankel covariance matrix have the same linear dependences as the rows of $H_{1,\infty}[K]$. This observation has led to the canonical correlation analysis proposed by Akaike [55]. A similar method based on rank tests of covariance matrices has been proposed in [12]. For deterministic I/O models, on the other hand, structure estimation methods have been proposed based on rank tests of the product-moment matrix of the input and output data [10, 11]. These methods will also work when the inputs and outputs are measured with noise, provided that the noise is white or its statistics are known. In [56] another method is proposed that applies to least squares models, that is, H(z) = I in (1) (see also [57]).

C. STRUCTURE ESTIMATION METHODS USING PSEUDOCANONICAL FORMS

When pseudocanonical forms are used, it is in principle only necessary to estimate the order n of the system. Result 2 summarizes the relevant theoretical results that are available.

Once the order has been estimated, say $\hat{n}$, any partition μ such that $|\mu| = \hat{n}$ can normally be used to obtain an identifiable model structure. It might well be, however, that the true system is close to the boundary of the selected neighborhood: that is, the coordinates could well be ill conditioned. For numerical reasons, therefore, it may be worth selecting, if not the best, at least a well-conditioned parametrization among the finite set of admissible ones (considering $\hat{n}$ as fixed). A number of methods have been proposed. Ljung and Rissanen have proposed a method based on the complexity of various submatrices of $R_{\hat{Y}}$, where $R_{\hat{Y}}$ is the covariance matrix of a vector $\hat{Y}_t^N$ made up of a finite set of predictors $\hat{y}(t + 1|t), \ldots, \hat{y}(t + N|t)$ [15]. Wertz, Gevers, and Hannan have proposed a Q - R factorization of Akaike's covariance matrix between future and past outputs, where at each step of the factorization the most independent remaining vector is added to the basis [17]. Van Overbeek and Ljung have proposed a method that is based on the conditioning of the information matrix; their procedure is not to search a priori for the "best" coordinate system, but to perform a coordinate transformation if the parameters in the present coordinate system become ill conditioned [16]. All these methods apply only to state-space models; in addition, they are covariance methods, which will only work when no deterministic inputs are present.

A lot more should be said about structure estimation. In particular, we have hardly touched on the consistency results for $\hat{K}_N(z)$ (or for $\hat{\tau}_N$) when the model order is either larger or smaller than that of the true system. We refer to the work of Hannan, Deistler, and Kavalieris for a discussion of these issues. A very readable discussion can be found in [58].

VIII. CONCLUSIONS

We have given a broad overview of the issues involved in selecting identifiable parametrizations and have presented most of the commonly used parametrizations and the techniques for selecting them. In the past decade researchers in this field have gained a much better understanding of the structure of multivariable systems, and yet there is still no consensus on a universal technique for the representation of such systems. This is due to several reasons. First there is still no agreement on a universal order or structure selection criterion, and there might never be one, because it is recognized more and more that an order or structure selection criterion will always have to incorporate a degree of subjectivity. The research of the past few years has focused on finding structure selection criteria that converge to the true structure when a true system is assumed to exist. Those "optimal" criteria require an enormous amount of computations, as they require the maximization of a large number of likelihood functions. This has led people to search for computationally cheaper suboptimal methods, most of which rely on very heuristic arguments. This is a major reason for the wide range of existing methods.

A second reason is that most researchers now recognize the fact that in most practical applications the true system being identified is infinite dimensional. It is not certain that an "optimal" criterion that converges to the true structure when such exists will also be the best one in the more realistic situation when no true system exists. In this situation the structure selection criterion should most certainly incorporate the intended use of the model. Very little if no effort has been made in this direction.

Because of this lack of universal agreement, the methods people use will be influenced very much by their familiarity with a particular method, by the intended use of the model, and by the availability of a particular software package. All of this probably accounts for the relatively small number of successful applications reported in the literature and the difficulty of comparing results obtained with different methods.

Finally, we wish to conclude on two practical notes. First, there is no denying that tremendous new insights have been gained from the theoretical research of the past decade on the estimation of structure of multivariable systems. However, since the generic parametrization is able to represent almost all systems (see Comment 5), in practice it is most often sufficient to estimate the order of the system, rather than its entire structure. Second, companion (or bloc-companion) forms are notoriously sensitive to numerical errors in the parameters. From a practical point of view, more research should be spent on finding numerically insensitive multivariable parametrizations

REFERENCES

1. R. E. KALMAN, "Algebraic Geometric Description of the Class of Linear Systems of Constant Dimension," *Annu. Princeton Conf. Inf. Sci. Syst., 8th, Princeton, New Jersey* (1974).

2. J. M. C. CLARK, "The Consistent Selection of Parametrizations in Systems Identification," *Joint Automatic Control Conf.*, Purdue University (1976).

3. W. DUNSMUIR and E. J. HANNAN, "Vector Linear Time-Series Models," *Adv. Appl. Probab. 8*, 339-364 (1976).

4. M. HAZEWINKEL, "Moduli and Canonical Forms for Linear Dynamical Systems II: The Topological Case," *Math. Syst. Theory 10*, 363-385 (1977).

5. D. G. LUENBERGER, "Canonical Forms for Linear Multivariable Systems," *IEEE Trans. Autom. Control AC-12*, 290-293 (1976).

6. D. Q. MAYNE, "A Canonical Model for Identification of Multivariable Linear Systems," *IEEE Trans. Autom. Control AC-17*, 728-729 (1972).

7. M. J. DENHAM, "Canonical Forms for the Identification of Multivariable Linear Systems," *IEEE Trans. Autom. Control AC-19*, 646-656 (1974).

8. K. GLOVER and J. C. WILLEMS, "Parametrization of Linear Dynamical Control Systems, Canonical Forms and Identifiability," *IEEE Trans. Autom. Control AC-19*, 640-646 (1974).

9. J. RISSANEN, "Basis of Invariants and Canonical Forms for Linear Dynamical Systems," *Automatica 10*, 175-182 (1974).

10. R. P. GUIDORZI, "Canonical Structures in the Identification of Multivariable Systems," *Automatica 11*, 361-374 (1975).

11. R. P. GUIDORZI, "Invariants and Canonical Forms for Systems Structural and Parametric Identification," *Atuomatica 17*, 117-133 (1981).

12. E. TSE and H. L. WEINERT, "Structure Determination and Parameter Identification for Multivariable Stochastic Linear Systems," *IEEE Trans. Autom. Control AC-20*, 603-613 (1975).

13. P. E. CAINES and J. RISSANEN, "Maximum Likelihood Estimation of Parameters in Multivariate Gaussian Stochastic Processes," *IEEE Trans. Inf. Theory IT-20*, 102-104 (1974).

14. M. HAZEWINKEL and R. E. KALMAN, "On Invariants, Canonical Forms and Moduli for Linear Constant Finite-Dimensional Dynamical Systems," *in* "Lecture Notes Economical-Mathematical System Theory, Vol. 131, pp. 48-60, Springer-Verlag, Berlin and New York, 1976.

15. L. LJUNG and J. RISSANEN, "On Canonical Forms, Parameter Identifiability and the Concept of Complexity," *Proc. IFAC Symp. Identif. Syst. Parameter Estim., 4th, Tbilisi*, 58-69 (1976).

16. A. J. M. VAN OVERBEEK and L. LJUNG, "On-Line Structure Selection for Multivariable State-Space Models," *Automatica 18*, 529-544 (1982).

17. V. WERTZ, M. GEVERS, and E. J. HANNAN, "The Determination of Optimum Structures for the State-Space Representation of Multivariate Stochastic Processes," *IEEE Trans. Autom. Control AC-27*, 1200-1211 (1982).

18. G. PICCI, "Some Numerical Aspects of Multivariable Systems Identification," *Math. Program. Stud. 18*, 76-101 (1982).

19. M. DEISTLER and E. J. HANNAN, "Some Properties of the Parametrization of ARMA Systems with Unknown Order," *J. Multivariate Anal. 11*, 474-484 (1981).

20. M. DEISTLER, "The Properties of the Parametrization of ARMAX Systems and Their Relevance for Structural Estimation," *Econometrica 51*, 1187-1207 (1983).

21. R. P. GUIDORZI and S. BEGHELLI, "Input-Output Multistructural Models in Multivariable Systems Identifaction," *Proc. IFAC Symp. Identif. Syst. Parameter Estim., 6th, Washington D.C.*, 461-466 (1982).

22. M. GEVERS and V. WERTZ, "Overlapping Parametrizations for the Representation of Multivariate Stationary Time Series," *in* "Geometry and Identification," pp. 73-99 (P. E. Caines and R. Hermann, eds.), Math. Sci. Press, Massachusetts, 1983.

23. M. GEVERS and V. WERTZ, "Uniquely Identifiable State-Space and ARMA Parametrizations for Multivariable Linear Systems," *Automatica 20*, 333-348 (1984).

24. G. O. CORREA and K. GLOVER, "Pseudocanonical Forms, Identifiable Parametrizations and Simple Parameter Estimation for Linear Multivariable Systems; Part I: Input-Output Models; Part II: Parameter Estimation," *Automatica 20*, 443-452 (1984).

25. M. DEISTLER, "General Structure and Parametrization of ARMA and State Space Systems and Its Relation to Statistical Problems," *in* "Handbook of Statistics," Vol. 5 (E. J. Hannan, P. R. Krishnaiah, and M. M. Rao, eds.), North Holland Publ., Amsterdam, 1983.

26. E. J. HANNAN and L. KAVALIERIS, "Multivariate Linear Time Series Models," *Adv. Appl. Probab. 16*, 492-561 (1984).

27. E. J. HANNAN, "The Estimation of the Order of an ARMA Process," *Ann. Stat. 8*, 1071-1081 (1980).

28. E. J. HANNAN, "Estimating the Dimension of a Linear System," *J. Multivariate Anal. 11*, 459-473 (1981).

29. T. KAILATH, "Linear Systems," Prentice-Hall, Englewood Cliffs, New Hersey, 1980.

30. W. A. WOLOWICH and H. ELLIOTT, "Discrete Models for Linear Multivariable Systems," *Int. J. Control 38*, 337-357 (1983).

31. M. GEVERS, "On ARMA Models, Their Kronecker Indices and Their McMillan Degree," *Int. J. Control 43*, 1745-1761 (1986)

32. L. LJUNG, "System Identification: Theory for the User," Prentice-Hall, Englewood Cliffs, New Jersey, 1986. 1986.

33. R. BELLMAN and K. J. ÅSTRÖM, "On Structural Identifiability, *Math. Biosci. 7*, 329-339 (1970).

34. L. LJUNG, "On Consistency and Identifiability," *Math. Program. Stud. 5*, 169-190 (1976).

35. V. SOLO, "Topics in Advanced Time Series Analysis," *in* "Lecture Notes in Mathematics," Springer-Verlag, Berlin and New York, 1983.

36. M. DEISTLER and H. G. SEIFERT, "Identifiability and Consistent Estimability in Econometric Models," *Econometrica 46*, 969-980 (1978).

37. V. M. POPOV, "Invariant Description of Linear, Time Invariant Controllable Systems," *SIAM J. Control 10*, 252-264 (1972).

38. A. C. ANTOULAS, "On Canonical Forms for Linear Constant Systems," *Int. J. Control 33*, 95-122 (1981).

39. W. A. WOLOWICH, "Linear Multivariable Systems," *Appl. Math. Sci. 11* (1974).

40. G. O. CORREA and K. GLOVER, "Multivariable Identification Using Pseudo-Canonical Forms," *Proc. IFAC Symp. Identif. Syst. Parameter Estim., 6th, Washington D.C.*, 1110-1115 (1982).

41. M. GEVERS and V. WERTZ, "On the Problem of Structure Selection for the Identification of Stationary Stochastic Processes," *Proc. IFAC Symp. Identif. Syst. Parameter Estim., 6th, Washington D.C.*, 287-292 (1982).

42. E. J. HANNAN, "The Identification and Parametrization of ARMAX and State Space Forms," *Econometrica 44*, 713-723 (1976).

43. P. STOICA, "On Full Matrix Fraction Descriptions," *Bull. Inst. Politech. Buc. Ser. Electro.*, 99-104 (1982).

44. M. DEISTLER, W. DUNSMUIR, and E. J. HANNAN, "Vector Linear Time Series Models: Corrections and Extensions," *Adv. Appl. Probab. 10*, 360-372 (1978).

45. L. KEVICZKY, J. BOKOR, and CS. BANYASZ, "A New Identification Method with Special Parametrization for Model Structure Determination," *Proc. IFAC Symp. Identif. Syst. Paramater Estim., 5th, Darmstadt, Federal Republic of Germany*, 561-568 (1979).

46. J. BOKOR and L. KEVICZKY, "Structure and Parameter Estimation of MIMO Systems Using Elementary Sub-System Representations," *Int. J. Control 39*, 965-986 (1984).

47. H. AKAIKE, "Information Theory and an Extension of the Maximum Likelihood Principle," *Proc. Int. Symp. Info. Theory, 2nd*; Supplement to "Problems of Control and Information Theory," 267-281 (1972).

48. H. AKAIKE, "On Entropy Maximisation Principle," *in* "Applications of Statistics," (P. R. Krishnaiah, ed.), North-Holland Publ., Amsterdam, 1977.

49. J. RISSANEN, "Modeling by Shortest Data Description," *Automatica 14*, 465-471 (1978).

50. J. RISSANEN, "Estimation of Structure by Minimum Description Length," *Circuits Syst. Signal Process. 1*, 395-406 (1982).

51. R. SHIBATA, "Asymptotically Efficient Selection of the Order of the Model for Estimating Parameters of a Linear Process," *Ann. Stat. 8*, 147-164 (1980).

52. R. SHIBATA, "An Optimal Autoregression Spectral Estimate," *Ann. Stat. 9*, 300-306 (1981).

53. L. KAVALIERIS, "Statistical Problems in Linear Systems," Ph.D. Dissertation, Australian National University, 1984.

54. J. DURBIN, "Fitting of Time Series Models," *Rev. Inst. Stat. 28*, 233-244 (1961).

55. H. AKAIKE, "Canonical Correlation Analysis of Time Series and the Use of an Information Criterion," *in* "System Identification: Advances and Case Studies," pp. 27-96 (R. K. Mehra and D. G. Lainiotis, eds.), Academic Press, New York, 1976.

56. R. P. GUIDORZI, M. LOSITO, and T. MURATORI, "The Range Error Test in the Structural Identification of Linear Multivariable Systems," *IEEE Trans. Autom. Control Ac-27*, 1044-1054 (1982).

57. P. STOICA, "Comments on 'The Range Error Test in the Structural Identification of Linear Multivariable Systems'," *IEEE Trans. Autom. Control AC-29*, 379-381 (1984).

58. M. DEISTLER, "Multivariate Time Series and Linear Dynamic Systems," *Adv. Stat. Methods Comput. 1*, in press (1985).

Parametric Methods for Identification of Transfer Functions of Linear Systems

LENNART LJUNG

Division of Automatic Control
Department of Electrical Engineering
Linköping University
Linköping, Sweden

I. INTRODUCTION

The identification of linear models of dynamical systems is of prime importance in many control and signal processing applications. In this contribution we shall assume that there is true linear systems S that generates the observed data. If y(t) and u(t) denote the output and the input, respectively, at time instant t, we assume that

$$y(t) = G_0(q)u(t) + v_0(t). \tag{1}$$

Here $G_0(q)$ is the transfer operator

$$G_0(q)u(t) = \left[\sum_{k=1}^{\infty} g_0(k)q^{-k}\right]u(t) = \sum_{k=1}^{\infty} g_0(k)u(t - k) \tag{2}$$

in the shift operator q $[qu(t) = u(t + 1); q^{-1}u(t) = u(t - 1)]$. We thus describe the system in discrete time, and for simplicity, the sampling interval is taken to be one time unit. In (1), $v_0(t)$ is an additive disturbance, which is supposed to be a stationary stochastic process with spectrum

$$\Phi_v(\omega) = \lambda_0|H_0(e^{i\omega})|^2. \tag{3}$$

This means that $\{v_0(t)\}$ can be regarded as generated by

$$v_0(t) = H_0(q)e_0(t), \tag{4}$$

where $\{e_0(t)\}$ is white noise with variance λ_0.

For the system (1) we may generate an input $\{u(t)\}$, possibly by output feedback as

$$u(t) = -F(q)y(t) + w(t), \tag{5}$$

where $\{w(t)\}$ is a deterministic sequence such that

$$\lim_{N\to\infty} \frac{1}{N} \sum_{t=1}^{N} w(t)w(t - \tau) = R_w(\tau)$$

exist for all τ, and

$$\Phi_w(\omega) = \sum_{\tau=-\infty}^{\infty} R_w(\tau)e^{-i\tau\omega}. \tag{6}$$

Thus by collecting the data set

$$z^N = \{u(1), y(1), \ldots, u(N), y(N)\}, \tag{7}$$

we may proceed to estimate the transfer functions G_0 and H_0 in (1), (4). Let the result be denoted by

$$\hat{G}_N(q) \ [= \hat{G}(q, z^N)] \quad \text{and} \quad \hat{H}_N(q) \ [= \hat{H}(q, z^N)]. \tag{8}$$

We shall discuss procedures for this in Section III.

Our objective is, of course, that the estimates (8) should be as close as possible to the true description. In Section II we shall discuss such model quality measures. In our search for good estimates we have several design variables available. We have already mentioned the feedback law $F(q)$ and the spectrum $\Phi_w(\omega)$. Other design variables will be associated with the estimation procedure. The main body of this contribution is devoted to a discussion of good or "optimal" choices of such design variables.

Now, estimation of transfer functions is a subject that has been widely discussed in the literature (see, e.g., [1], [2], [3], and [4]). The particular approach taken here follows the development in [5] through [11].

II. MEASURES OF MODEL QUALITY

A. *THE TRUE SYSTEM AND THE MODEL*

Suppose that the true system is subject to (1) through (4), that is, that

$$y(t) = G_0(q)u(t) + H_0(q)e_0(t), \tag{9}$$

where $\{e_0(t)\}$ is white noise with variance λ_0.

For simpler notation, we shall also use

$$T_0(q) = [G_0(q)\ H_0(q)]. \tag{10}$$

Suppose that we have decided upon all the design variables $\mathscr{D}$ and as a result obtained the model

$$\hat{T}(q, \mathscr{D}) = \left[\hat{G}(q, \mathscr{D})\hat{H}(q, \mathscr{D})\right]. \tag{11}$$

Here $\mathscr{D}$ will contain, among other things, N, the number of collected data.

B. *THE INTENDED USE OF THE MODEL*

It is, of course, desirable that the model $T(q, \mathscr{D})$ be close to $T_0(q)$. The difference

$$\tilde{T}(e^{i\omega}, \mathscr{D}) \triangleq \hat{T}(e^{i\omega}, \mathscr{D}) - T_0(e^{i\omega}) \tag{12}$$

should, in other words, be small. Now, it may be unnecessary to demand that $\tilde{T}$ be small for all frequencies ω. Depending on the intended use of the model, some frequency ranges may be more important than others.

This idea could be formulated as follows. Let s(t) be a signal derived from a model application. It could, for example, be the output of the system when a minimum variance regulator, computed using the model, is applied, or the predicted output of the system using a predictor based on the model or something else. We shall give specific examples shortly. Conceptually, we could write

$$s(t) = f(T(q))w(t) \tag{13}$$

to denote that the transfer function T, as well as some additional signals w(t) (reference signals and/or noises), are used to determine s(t). The true system's properties would then be contained in the function f.

$$s_0(t) = f(T_0(q))w(t). \tag{14}$$

When instead the model (11) is used, we get the result

$$\hat{s}_{\mathscr{D}}(t) = f(\hat{T}(q, \mathscr{D}))w(t). \tag{15}$$

It is now of interest to evaluate the *performance degradation*

$$\tilde{s}_{\mathscr{D}}(t) \triangleq \hat{s}_{\mathscr{D}}(t) - s_0(t). \tag{16}$$

When the error (12) is small, we could use Taylor's expansion to derive the approximate expression

$$\tilde{s}_{\mathscr{D}}(t) \approx \tilde{T}(q, \mathscr{D})F(q)w(t), \tag{17}$$

where

$$F(q) = \frac{\partial}{\partial T} f(T)\Big|_{T=T_0(q)} \qquad (2 \times r \text{ matrix;} \quad r = \dim w). \tag{18}$$

The spectrum of $\tilde{s}_{\mathscr{D}}(t)$ then is

$$\begin{aligned}\tilde{\Phi}_s(\omega, \mathscr{D}) &= T(e^{i\omega}, \mathscr{D})F(e^{i\omega})\Phi_w(\omega)F^T(e^{-i\omega})T^T(e^{-i\omega}, \mathscr{D}) \\ &= \operatorname{tr}\Big[\overline{S}(\omega, \mathscr{D}) \cdot C_1'(\omega)\Big],\end{aligned} \tag{19}$$

where

$$\tilde{S}(\omega, \mathscr{D}) = \tilde{T}^T(e^{i\omega}, \mathscr{D})\tilde{T}(e^{-i\omega}, \mathscr{D}) \quad (2 \times 2 \text{ matrix}) \tag{20}$$

and

$$C_1'(\omega) = F(e^{i\omega})\Phi_w(\omega)F^T(e^{-i\omega}) \quad (2 \times 2 \text{ matrix}). \tag{21}$$

The spectrum $\tilde{\Phi}_s(\omega, \mathscr{D})$ is a random process, since the matrix $\tilde{S}(\omega, \mathscr{D})$ depends on the transfer function estimate $\hat{T}(e^{i\omega}, \mathscr{D})$, which is a random vector. To obtain a deterministic measure of the performance degradation $\tilde{s}_{\mathscr{D}}(t)$, we could thus consider the expected value

$$\Psi_s(\omega, \mathscr{D}) = E\tilde{\Phi}_s(\omega, \mathscr{D}). \tag{22}$$

This expression gives us insight into the character of the performance degradation over the frequency domain. We could evaluate the variance of the error signal $\tilde{s}_{\mathscr{D}}(t)$ by

$$E\tilde{s}_{\mathscr{D}}^2(t) = \frac{1}{2\pi}\int_{-\pi}^{\pi} \Psi_s(\omega, \mathscr{D})\, d\omega. \tag{23}$$

We could also consider a weighted frequency domain norm of the error spectrum:

$$J(\mathscr{D}) = \int_{-\pi}^{\pi} \Psi_s(\omega, \mathscr{D})\alpha(\omega)\, d\omega, \tag{24}$$

where $\alpha(\omega)$ is a weighting function that reflects the importance of having a small error spectrum at certain frequencies.

C. A DESIGN VARIABLE CRITERION

By defining $C(\omega) = \alpha(\omega)C_1'(\omega)$, we have thus been lead to an objective criterion

$$J(\mathscr{D}) = \int_{-\pi}^{\pi} \Psi_s(\omega, \mathscr{D})\alpha(\omega)\, d\omega = \int_{-\pi}^{\pi} \mathrm{tr}[\overline{\Pi}(\omega, \mathscr{D})C(\omega)]\, d\omega, \tag{25}$$

where

$$\Pi(\omega, \mathscr{D}) = E\tilde{T}^T(e^{i\omega}, \mathscr{D})\tilde{T}(e^{-i\omega}, \mathscr{D}) \tag{26}$$

and

$$C(\omega) = \begin{pmatrix} C_{11}(\omega) & C_{12}(\omega) \\ C_{21}(\omega) & C_{22}(\omega) \end{pmatrix} \tag{27}$$

reflects our relative interest in different frequency bands. The problem of choosing design variables can now be stated as

$$\min_{\mathscr{D}\in\Delta} J(\mathscr{D}), \tag{28}$$

where Δ denotes the constraints associated with our desire to do at most "a reasonable amount of work." These will typically include a maximum number of samples, signal power constraints, not too complex numerical procedures, and so on. The constraints Δ could also include that certain design variables simply are not available to the user in the particular application in question.

Let us give some examples of model applications that lead to different functions $C(\omega)$ in (27).

D. *MODEL APPLICATIONS*

Example 2.1: Simulation. Suppose that we use our model to simulate the output with an input with spectrum $\Phi_u^*(\omega)$ (we use superscript asterisk to distinguish this input spectrum from the one used during the identification experiment). That gives us an output (for simplicity we suppress the argument $\mathscr{D}$ in $\hat{G}$)

$$\hat{y}_{\mathscr{D}}(t) = \hat{G}(q)u^*(t), \tag{29}$$

which differs from the one the true system $G_0(q)$ would give by

$$\tilde{y}(t) = \tilde{G}(q)u^*(t).$$

This difference has the spectrum

$$\Phi_{\tilde{y}}(\omega) = |\tilde{G}(e^{i\omega})|^2 \Phi_u^*(\omega). \tag{30}$$

We thus have the situation described by (19) with

$$C(\omega) = \begin{pmatrix} \Phi_u^*(\omega) & 0 \\ 0 & 0 \end{pmatrix}. \tag{31}$$

Example 2.2: Prediction. The one-step-ahead prediction error when using the current model is

$$\hat{\epsilon}_{\mathscr{D}}(t) = \hat{H}^{-1}(q)y(t) - \hat{H}^{-1}(q)\hat{G}(q)u(t). \tag{32}$$

The difference to the error obtained for the true system e(t) obtained as

$$\begin{aligned} \tilde{\epsilon}_{\mathscr{D}}(t) &= \hat{\epsilon}_{\mathscr{D}}(t) - e(t) \\ &= -\frac{1}{\hat{H}}\left[\left(G - G_0^*\right)u(t) + \left(\hat{H} - H_0\right)e(t)\right], \end{aligned} \tag{33}$$

where we suppressed arguments.

Now use the approximation $\hat{H} \approx H_0$. Then

$$\tilde{\epsilon}_{\mathscr{D}}(t) = -(1/H_0)(\tilde{G}u + \tilde{H}e).$$

This means that in (17) through (19) we take

$$C(\omega) = \frac{1}{|H_0(e^{i\omega})|^2}\begin{pmatrix} \Phi_u^*(\omega) & \Phi_{ue}^*(\omega) \\ \Phi_{ue}^*(-\omega) & \lambda^* \end{pmatrix}, \tag{34}$$

where superscript asterisk indicates that these are the spectra for the input data to which the predictor is applied (very possibly different from those during the identification experiment).

Example 2.3: Generalized Minimum Variance Control. Suppose that our control objective is to obtain a closed loop system

$$y(t) = R(q)e(t). \tag{35}$$

The case $R(q) = 1$ corresponds to the well-known minimum variance control [12], but in practice it could be worthwhile to aim at (35) to obtain a better trade-off between input energy and output variance.

It is easy to verify that the control law

$$u(t) = \frac{R(q) - \hat{H}(q)}{\hat{G}(q)R(q)} y(t) \tag{36}$$

would lead to (35) in case the system was described by the model $\hat{G}$, $\hat{H}$. This is a reasonable regulator only if $\hat{G}(q)$ has no zeros outside the unit circle. Inserting (36) into the true system (9) gives, after some calculations,

$$y_{\mathscr{D}}(t) \simeq R(q)e(t) + \frac{R(q)}{H_0(q)}\left[\frac{H_0(q) - R(q)}{G_0(q)} \tilde{G}(q) - \tilde{H}(q)\right]e(t), \tag{37}$$

assuming both $\tilde{G}$, $\tilde{H}$ to be small. In terms of (17) through (19), we thus have

$$C(\omega) = \lambda \left|\frac{R(e^{i\omega})}{H_0(e^{i\omega})}\right|^2$$

$$\times \begin{bmatrix} \dfrac{|H_0(e^{i\omega}) - R(e^{i\omega})|^2}{|G_0(e^{i\omega})|^2} & -\dfrac{H_0(e^{i\omega}) - R(e^{i\omega})}{G_0(e^{i\omega})} \\ -\dfrac{H_0(e^{-i\omega}) - R(e^{-i\omega})}{G_0(e^{-i\omega})} & 1 \end{bmatrix}. \tag{38}$$

Example 2.4: Pole Placement. Now, assume that we want to create a servo system from a reference input $r(t)$ to the output, such that

$$y(t) = R(q)r(t) \tag{39}$$

and such that the noise spectrum is insignificant in comparison with the spectrum of r, $\Phi_r(\omega)$. A general solution to obtain

(39), in case G has no zeros outside the unit circle, is to use the regulator

$$u(t) = Q(q)r(t) - P(q)y(t) \tag{40}$$

with transfer functions Q and P such that

$$\frac{\hat{G}(q)Q(q)}{1 + \hat{G}(q)P(q)} = R(q). \tag{41}$$

Using (40) in the true system gives

$$y_{\mathscr{D}}(t) = \frac{G_0 Q}{1 + G_0 P} r(t) = Rr + \left[\frac{G_0 Q}{1 + G_0 P} - \frac{\hat{G}Q}{1 + \hat{G}P}\right] r$$

$$\simeq Rr + \frac{Q\tilde{G}}{(1 + G_0 P)^2} r$$

if $\tilde{G}$ is small. Comparing with (17) through (19), we see that this corresponds to

$$C(\omega) = \begin{bmatrix} \dfrac{|R|^2}{|G_0|^2 |1 + G_0 P|^2} \Phi_r(\omega) & 0 \\ 0 & 0 \end{bmatrix}. \tag{42}$$

III. PREDICTION ERROR IDENTIFICATION METHODS

A. *THE MODEL SET*

The perhaps most common approach in modern identification is to postulate that the transfer function is to be sought within a certain set:

$$\mathscr{G} = \left\{G(e^{i\omega}, \theta) \mid \theta \in D_{\mathscr{M}}\right\}. \tag{43}$$

Here $D_{\mathscr{M}}$ typically is a subset of R^d. In order to improve the result, it is customary to also include assumptions about the disturbance spectrum $\Phi_v(\omega)$ [see (1)-(7)]. It is assumed to belong to a set

$$\Phi_v(\omega) = \lambda|H(e^{i\omega}, \Theta)|^2; \quad H(e^{i\omega}, \Theta) \in \mathscr{H}, \tag{44}$$

$$\mathscr{H} = \left\{H(e^{i\omega}, \Theta) \mid \Theta \in D_{\mathscr{M}}\right\}.$$

This means that the system is assumed to be described as

$$y(t) = G(q, \Theta)u(t) + H(q, \Theta)e(t) \tag{45}$$

for some $\Theta \in D_{\mathscr{M}}$. Here $\{e(t)\}$ is a sequence of independent random variables with zero mean values and variances λ and G and H are functions of the shift operator q;

$$G(q, \Theta) = \sum_{k=1}^{\infty} g_k(\Theta)q^{-k}, \tag{46a}$$

$$H(q, \Theta) = 1 + \sum_{k=1}^{\infty} h_k(\Theta)q^{-k}. \tag{46b}$$

There are several ways by which the transfer functions in (45) can be parametrized. Two common ones are illustrated in the following examples.

Example 3.1: State-Space Models. Suppose that a model of the system is posed in state-space, innovations form:

$$\begin{aligned} x(t + 1) &= A(\Theta)x(t) + B(\Theta)u(t) + K(\Theta)e(t), \\ y(t) &= C(\Theta)x(t) + e(t). \end{aligned} \tag{47}$$

Here the matrices A, B, C, and K may be parametrized by Θ in an arbitrary way. This model corresponds to (55) with

$$\begin{aligned} G(q, \Theta) &= C(\Theta)[qI - A(\Theta)]^{-1}B(\Theta), \\ H(q, \Theta) &= 1 + C(\Theta)[qI - A(\Theta)]^{-1}K(\Theta). \blacksquare \end{aligned} \tag{48}$$

Example 3.2: ARMAX Models. Suppose that the model is chosen as

$$A(q^{-1})y(t) = B(q^{-1})u(t) + C(q^{-1})e(t), \tag{49}$$

where A, B, and C are polynomials in the delay operator. Such a model is known as an ARMAX model. Clearly (49) corresponds to (45) with

$$G(q, \Theta) = B(q^{-1})/A(q^{-1}), \tag{50a}$$

$$H(q, \Theta) = C(q^{-1})/A(q^{-1}). \tag{50b}$$

Here the parameter vector Θ consists of the coefficients of the polynomials A, B, and C. ■

B. THE ESTIMATION METHOD

Given the model (45) and input-output data up to time t - 1, we can determine the predicted output at time t as follows. Rewrite (45) as

$$H^{-1}(q, \Theta)y(t) = H^{-1}(q, \Theta)G(q, \Theta)u(t) + e(t)$$

or

$$y(t) = [1 - H^{-1}(q, \Theta)]y(t) + H^{-1}(q, \Theta)G(q, \Theta)u(t) + e(t).$$

Since $H^{-1}(q, \Theta)$ has an expansion in powers of q^{-1} that starts with a "one" [see (46b)] and G(q, Θ) contains a delay, the right-hand side of the preceding expression contains y(k) and u(k) only for $k \leq t - 1$. The term e(t) is independent of everything that happened up to time t - 1. Hence the natural one-step-ahead prediction, equal to the conditional expectation of y(t) given previous data, is

$$\hat{y}(t \mid \Theta) = (1 - H^{-1}(q, \Theta))y(t) + H^{-1}(q, \Theta)G(q, \Theta)u(t). \tag{51}$$

At time t, when y(t) has been recorded, we can compute the prediction error that the model (45) led to

$$\epsilon(t, \Theta) = y(t) - \hat{y}(t \mid \Theta) = H^{-1}(q, \Theta)(y(t) - G(q, \Theta)u(t)). \tag{52}$$

We may say that the model (45) is "good" if the sequence $\epsilon(t, \Theta)$, $t = 1, 2, \ldots, N$, is "small." In a very common class of identification methods, the squared sum of prediction errors is minimized to find the "best" model:

$$\hat{\Theta}_N = \arg \min_{\Theta \in D_{\mathcal{M}}} \frac{1}{N} \sum_{t=1}^{N} \epsilon^2(t, \Theta). \tag{53}$$

With $\hat{\Theta}_N$ determined in this way, the transfer function estimate becomes

$$\hat{G}_N(e^{i\omega}) = G\left(e^{i\omega}, \hat{\Theta}_N\right). \tag{54}$$

Among methods that can be expressed as (53) we find the maximum likelihood method (assuming Gaussian disturbances), the least squares method, and others. See [13] and [14] for a further discussion.

C. *SOME EXTENSIONS*

It may often be worthwhile to consider a modified criterion (53) where the prediction errors $\epsilon(t, \Theta)$ (or equivalently the input-output sequences) first are filtered through a filter $L(q)$:

$$\epsilon_F(t, \Theta) = L(q)\epsilon(t, \Theta). \tag{55}$$

This is, however, equivalent to replacing the noise model $H(q, \Theta)$ by $H(q, \Theta)/L(q)$. See (52) and for a further discussion [7]. Prefiltering the data thus corresponds to selecting another noise model set.

Also the use of k-step-ahead predictors in (51) might be useful. As elaborated on in [7], k-step-ahead prediction methods are equivalent to replacing $H(q, \Theta)$ by

$$H(q, \Theta)M_k^{-1}(q, \Theta), \tag{56}$$

where $M_k(q, \Theta)$ are the first k terms in the Laurent expansion of $H(q, \Theta)$. The use of k-step-ahead predictors is thus equivalent to prefiltering $[L(q) = M_k(q, \Theta)]$ or to selecting another noise model set.

D. DESIGN VARIABLES

Let us list the available design variables:

$$\Phi_w(\omega): \quad \text{spectrum of the extra input } w \text{ in (5)}, \tag{57a}$$

$$F(q): \quad \text{feedback law in (5)}, \tag{57b}$$

$$\mathscr{H} = \{H(q, \Theta) \mid \Theta \in D_{\mathscr{M}}\}: \quad \text{set of noise models.} \tag{57c}$$

The later variable includes, as we noted, the possibility of prefiltering with L in (55) and the use of k-step-ahead predictors [see (56)]. These three items will henceforth be denoted collectively by the symbol $\mathscr{D}$.

Other design variables, such as N, the number of collected data, and $\mathscr{G}$, the set of transfer function models (including the model order n), will be regarded as fixed in this study.

It will be useful to translate the design variables Φ_w and F to the resulting input spectrum Φ_u and cross spectrum Φ_{ue} between u and e in (9). Since (9) and (5) give the closed loop system,

$$y(t) = \frac{G(q)}{1 + F(q)G(q)} w(t) + \frac{H(q)}{1 + F(q)G(q)} e(t), \tag{58a}$$

$$u(t) = \frac{F(q)H(q)}{1 + F(q)G(q)} e(t) + \frac{1}{1 + F(q)G(q)} w(t). \tag{58b}$$

We find that

$$\Phi_u(\omega) = \lambda_0 \left| \frac{F(e^{i\omega})H(e^{i\omega})}{1 + F(e^{i\omega})G(e^{i\omega})} \right|^2 + \Phi_w(\omega) \left| \frac{1}{1 + F(e^{i\omega})G(e^{i\omega})} \right|^2,$$

$$\Phi_{ue}(\omega) = \frac{F(e^{i\omega})H(e^{i\omega})}{1 + F(e^{i\omega})G(e^{i\omega})} \lambda_0. \tag{59}$$

IV. ASYMPTOTIC PROPERTIES OF THE ESTIMATED TRANSFER FUNCTIONS

A. *CONVERGENCE*

Under weak conditions it can be shown that

$$\hat{\Theta}_N \to \Theta^* = \arg \min_{\Theta \in D_{\mathcal{M}}} \bar{V}(\Theta) \quad \text{w.p.1.} \quad \text{as} \quad N \to \infty, \tag{60}$$

where

$$\bar{V}(\Theta) = \lim_{N\to\infty} \frac{1}{N} \sum_{t=1}^{N} E\epsilon^2(t, \Theta). \tag{61}$$

(See, e.g., [13], [11].)

Applying Parseval's relationship to (61) gives

$$\bar{V}(\Theta) = \frac{1}{2\pi} \int_{-\pi}^{\pi} \Phi_\epsilon(\omega, \Theta)\, d\omega, \tag{62}$$

where Φ_ϵ is the spectrum of the prediction errors.

Now

$$\begin{aligned}
\epsilon(t, \Theta) &= H^{-1}(q, \Theta)[y(t) - G(q, \Theta)u(t)] \\
&= H^{-1}(q, \Theta)\left[-\tilde{G}(q, \Theta)u(t) + H_0(q)e_0(t)\right] \\
&= e_0(t) - H^{-1}(q, \Theta)\left[\tilde{G}(q, \Theta)u(t) + \tilde{H}(q, \Theta)e_0(t)\right] \\
&= e_0(t) - H^{-1}(q, \Theta)\tilde{T}(q, \Theta)\chi(t),
\end{aligned} \tag{63}$$

where

$$\begin{aligned}
\tilde{T}(q, \Theta) &= T(q, \Theta) - T_0(q), \\
T(q, \Theta) &= [G(q, \Theta)\ H(q, \Theta)]
\end{aligned} \tag{64}$$

and

$$\chi(t) = \begin{pmatrix} u(t) \\ e_0(t) \end{pmatrix}. \tag{65}$$

Hence

$$\Phi_{\epsilon}(\omega, \Theta) = \lambda_0 + \tilde{T}(e^{i\omega}, \Theta)\Phi_{\chi}(\omega)\tilde{T}^T(e^{-i\omega}, \Theta)/|H(e^{i\omega}, \Theta)|^2, \tag{66}$$

with

$$\Phi_{\chi}(\omega) = \begin{bmatrix} \Phi_u(\omega) & \Phi_{ue}(\omega) \\ \Phi_{ue}(-\omega) & \lambda_0 \end{bmatrix}. \tag{67}$$

Combining (60)-(62) and (66) thus gives the following characterization of the limit model:

$$\Theta^* = \arg\min_{\Theta} \int_{-\pi}^{\pi} \mathrm{tr}[R(\omega, \Theta) \cdot Q(\omega, \Theta)]\, d\omega, \tag{68}$$

with

$$R(\omega, \Theta) = \tilde{T}^T(e^{-i\omega}, \Theta)\tilde{T}(e^{i\omega}, \Theta), \tag{69}$$

$$Q(\omega, \Theta) = \Phi_{\chi}(\omega)/|H(e^{i\omega}, \Theta)|^2. \tag{70}$$

Note that for open loop operation $[\Phi_{ue}(\omega) \equiv 0]$ and a fixed (Θ-independent) noise model, $H(e^{i\omega}, \Theta) = H_*(e^{i\omega})$, this expression specializes to

$$\Theta^* = \arg\min_{\Theta} \int_{-\pi}^{\pi} |G(e^{i\omega}, \Theta)|^2 \cdot \frac{\Phi_u(\omega)}{|H_*(e^{i\omega})|^2}\, d\omega. \tag{71}$$

B. VARIANCE

Let

$$T^*(q) = T(q, \Theta^*), \tag{72}$$

with Θ^* defined as before.

Under fairly general conditions it can then be shown that

$$\sqrt{N}\left[\hat{T}_N(e^{i\omega}) - T^*(e^{i\omega})\right] \in \mathrm{AsN}(0, P_n(\omega)). \tag{73}$$

Here (73) means that the random variable on the left converges in distribution to the normal distribution with zero mean and

[We define the covariance of a complex valued random variable $\hat{T}$ as

$$\text{Cov } \hat{T} = E(\hat{T} - E\hat{T})\overline{(\hat{T} - E\hat{T})},$$

where overbar means complex conjugate.]

Results such as (73) go back to the asymptotic normality of the parameter estimate $\hat{\theta}_N$, established, for example, in [15]. The expression for $P_n(\omega)$ is in general complicated. For models that are parametrized as "black boxes," we have, however, the following general result [6]:

$$\lim_{n\to\infty} \frac{1}{n} P_n(\omega) = \Phi_v(\omega)[\Phi_\chi(\omega)]^{-1}, \tag{74}$$

with Φ_v and Φ_χ defined by (3) and (67), respectively and n is the model order.

C. A PRAGMATIC INTERPRETATION

Even though the covariance of $\hat{T}_N$ need not converge (convergence in distribution does not imply convergence in L_2), we allow ourselves to use the result (73)-(74) in the following more suggestive version:

$$\text{Cov } \hat{T}_N(e^{i\omega}) \approx (n/N)\Phi_v(\omega)[\Phi_\chi(\omega)]^{-1}. \tag{75}$$

We shall also sllow the approximation

$$E\hat{T}_N(e^{i\omega}) \approx T^*(e^{i\omega}). \tag{76}$$

(See [11] for justifications.)

With (75) and (76) the expression (26) can be rewritten

$$\begin{aligned} \Pi(\omega, \mathscr{D}) &= E\tilde{T}^T(e^{i\omega}, \mathscr{D})\tilde{T}(e^{-i\omega}, \mathscr{D}) \\ &= B(\omega, \mathscr{D}) + \text{Cov } \hat{T}_N(e^{i\omega}, \mathscr{D}), \end{aligned} \tag{77}$$

where the bias contribution is

$$\begin{aligned} B(\omega, \mathscr{D}) = R(\omega, \theta^*(\mathscr{D})) &= \left[T(e^{i\omega}, \theta^*(\mathscr{D})) - T_0(e^{i\omega})\right]^T \\ &\times \left[T(e^{-i\omega}, \theta^*(\mathscr{D})) - T_0(e^{-i\omega})\right]. \end{aligned} \tag{78}$$

The function R was defined in (69). We have here appended the argument $\mathscr{D}(\Theta^* = \Theta^*(\mathscr{D}),\ \hat{\Theta}_N = \hat{\Theta}_N(\mathscr{D}))$ to stress the dependence on the design variables.

The criterion (25) can thus be split into a bias and a variance contribution:

$$J(\mathscr{D}) \simeq J_B(\mathscr{D}) + J_P(\mathscr{D}), \tag{79}$$

where

$$J_B(\mathscr{D}) = \int_{-\pi}^{\pi} \operatorname{tr} B(\omega, \mathscr{D}) C(\omega)\, d\omega, \tag{80}$$

$$J_P(\mathscr{D}) = \int_{-\pi}^{\pi} \operatorname{tr}\left[\operatorname{Cov} \hat{T}_N(e^{i\omega}, \mathscr{D}) \cdot C(\omega)\, d\omega\right]$$

$$\simeq \frac{n}{N} \int_{-\pi}^{\pi} \Phi_v(\omega) \cdot \operatorname{tr}\left[\Phi_\chi^{-1}(\omega, \mathscr{D}) C(\omega)\right] d\omega. \tag{81}$$

In the following two sections we shall discuss the minimization of these two contributions to the design criterion.

V. MINIMIZING THE BIAS CONTRIBUTION

Consider now the problem of minimizing the bias distribution, that is,

$$\min_{\mathscr{D} \in \Delta} J_B(\mathscr{D}), \tag{82}$$

where $J_B(\mathscr{D})$ is defined by (78), (80). The function $J_B(\mathscr{D})$ depends on $\mathscr{D}$ via $\Theta^*(\mathscr{D})$. The dependence on $\mathscr{D}$ of the latter function, in turn, is defined by (68), which we write as

$$\Theta^*(\mathscr{D}) = \arg\min_{\Theta} \int_{-\pi}^{\pi} \operatorname{tr}[R(\omega, \Theta) \cdot Q(\omega, \Theta, \mathscr{D})]\, d\omega. \tag{83}$$

Here $R(\omega, \Theta)$ is defined by (69) and Q by (70). We have appended the argument $\mathscr{D}$ to Q to stress that it is made up from the design variables (57). (See also [11, [8], and [9].)

A. THE CASE OF A FIXED NOISE MODEL

Suppose first that $\mathscr{H}$ in (57c) is removed from the design variable and considered as a priori chosen to the fixed model $H_*(q)$:

$$\mathscr{H} = \{H_*(q)\}. \tag{84}$$

Then the function $Q(\omega, \Theta, \mathscr{D})$ in (70) does not depend on Θ:

$$Q(\omega, \Theta, \mathscr{D}) = Q_*(\omega, \mathscr{D}) = \Phi_\chi(\omega)/|H_*(e^{i\omega})|^2. \tag{85}$$

We also note that when $H_*(q)$ is fixed, $\Theta^*(\mathscr{D})$ does not depend on the (10) element of Q, that is, $\lambda_0/|H_*(e^{i\omega})|^2$. [This element meets the fixed function $|H_0(e^{i\omega}) - H_*(e^{i\omega})|^2$ in (68).]

Turning to optimization of (82), we define the function

$$W(\Theta, \bar{Q}(\omega)) = \int_{-\pi}^{\pi} \operatorname{tr} R(\omega, \Theta)\bar{Q}(\omega)\, d\omega. \tag{86}$$

Then, according to (83) and (85),

$$\Theta^*(\mathscr{D}) = \arg\min_{\Theta} W(\Theta, Q_*(\omega, \mathscr{D})). \tag{87}$$

Here the design variables $\mathscr{D}$ uniquely define the function $Q(\omega, \mathscr{D})$ above.

Similarly for (82),

$$J_B(\mathscr{D}) = W(\Theta^*(\mathscr{D}), C(\omega)). \tag{88}$$

At this point, the following lemma is useful.

Lemma 5.1. Let $V(x, y)$ be a scalar-valued function of two variables such that each may take values in some general Hilbert space. Let for fixed y

$$x^*(y) = \arg\min_{x} V(x, y), \tag{89}$$

and let for a fixed z

$$y^*(z) = \arg\min_{y} V(x^*(y), z), \tag{90}$$

assuming that these minimizing values are unique and well defined. Then

$$y^*(z) = z. \tag{91}$$

Proof. By the definition (89)

$$V(x^*(z),\ z) \le V(x,\ z) \quad \forall x,\ \forall z.$$

Hence

$$V(x^*(z),\ z) \le V(x^*(y),\ z) \quad \forall y. \tag{92}$$

From (90), by definition

$$V(x^*(y^*(z)),\ z) \le V(x^*(y),\ z) \quad \forall y. \tag{93}$$

Now (92) and (93) imply

$$x^*(y^*(z)) = x^*(z),$$

which implies (87) since the mapping x^* is assumed to be injective [follows from the assumed uniqueness of (90)]. ∎

Remark. The assumption of uniqueness in (90) can be relaxed by instead considering the *set*

$$y^*(z) = \left\{ y \mid V(x^*(y),\ z) = \min_y V(x^*(y),\ z) \right\}.$$

The statement corresponding to (91) then is

$$z \in y^*(z),$$

which is sufficient for our purposes. ∎

Applying the lemma to (87), (88), we find that

$$\min_{\mathscr{D}\in\Delta} J_B(\mathscr{D}) \Rightarrow \mathscr{D}_{opt} : Q_*(\omega,\ \mathscr{D}_{opt}) = \alpha \cdot C(\omega), \tag{94}$$

provided that $\alpha \cdot C(\omega)$ belongs to the admissible designs Δ for some positive scalar α. According to what we said earlier, the 2, 2 element of Q_* can be chosen freely when satisfying (94).

Consequently, the optimal design according to the criterion (82) is to select the input spectrum Φ_u and the cross spectrum

Φ_{ue}, so that

$$\frac{1}{|H_*(e^{i\omega})|^2}\begin{bmatrix}\Phi_u(\omega) & \Phi_{ue}(\omega) \\ \Phi_{ue}(-\omega) & *\end{bmatrix} = \alpha\begin{bmatrix}C_{11}(\omega) & C_{12}(\omega) \\ C_{21}(\omega) & *\end{bmatrix}. \tag{95}$$

Here the asterisk means that the element in question does not affect the optimal design.

This gives a concrete recipe for constructing optimal designs for bias minimization.

B. *THE NOISE MODEL SET IS A SINGLETON; PENALTY ON TRANSFER FUNCTION G ONLY*

Let us now consider the case

$$\mathscr{H} = [H_*(q)], \tag{96}$$

where H_* is a design variable; that is, we still confine the noise model to be fixed, but the choice of it is included in the design variables $\mathscr{D}$. Let us study the special case

$$C(\omega) = \begin{bmatrix}C_{11}(\omega) & 0 \\ 0 & 0\end{bmatrix}, \tag{97}$$

that is, where we penalize only the error in the transfer function estimate, $G_N(e^{i\omega}, \theta^*(\mathscr{D})) - G_0(e^{i\omega})$. (Examples 2.1 and 2.4 led to this special case.)

It is clear from (95) that in this case open loop operation $[\Phi_{ue}(\omega) \equiv 0]$ is optimal, regardless of H_*. Also (97) implies that the discrepancy $H_* - H_0$ does not affect the criterion (82). Hence H_* can be chosen freely, and (95) implies the following result. [Here we include the prefilter L in (55) as an explicit option.]

Theorem 5.1. Consider the problem to minimize (82) with respect to $\mathscr{D} = \left\{\Phi_u(\omega),\ \Phi_{ue}(\omega),\ L(e^{i\omega}),\ H_*(e^{i\omega})\right\}$ under the assumptions (96)-(97). Then $\mathscr{D}_{opt}$ is such that

$$\Phi_{ue}^{opt}(\omega) \equiv 0, \tag{98a}$$

$$\frac{|L(e^{i\omega})|^2 \Phi_u^{opt}(\omega)}{|H_*^{opt}(e^{i\omega})|^2} = \alpha \cdot C_{11}(\omega), \tag{98b}$$

where α is any constant that makes $\mathscr{D}_{opt}$ belong to the admissible set. ∎

Notice that there are several ways of obtaining the optimal design. Any combinations of input spectrum and noise model that obey (98b) will give the optimal bias distribution. Also recall that the choice of noise model $H_*(q)$ contains the option of prediction horizon k [see (56)].

C. THE GENERAL CASE

When the noise model set is parametrized by the parameter vector Θ, no formal results of the character (95) can be proved in the general case. The reason is that when Q in (84) depends on Θ, Lemma 5.1 cannot be applied. However, one may, of course, still use (95) in a pragmatic fashion by replacing

$$H_*(e^{i\omega}) \quad \text{by} \quad H(e^{i\omega}, \Theta^*(\mathscr{D})). \tag{99}$$

This makes (95) an equation for $\mathscr{D}_{opt}$. In the special case of (97) and independent parametrization of G and H [$\Theta = (\rho, \eta)^T$; $G(q, \Theta) = G(q, \rho)$, $H(q, \Theta) = H(q, \eta)$], the result (95) and (99) is still formally correct (see [11]). In other cases, (99) may still be reasonable, as we shall demonstrate subsequently.

D. SOME EXAMPLES

Example 5.1: Bias in Output Error Identification. Consider the fourth-order system

$$G_0(q) = \frac{0.001q^{-2}(10 + 7.4q^{-1} + 0.924q^{-2} + 0.1764q^{-3})}{1 - 2.14q^{-1} + 1.555q^{-2} - 0.4387q^{-3} + 0.04203q^{-4}}, \quad (100)$$

which has double poles in 0.82 and 0.25, a double zero at -0.42, and a single zero at -0.1.

A model

$$y(t) = \frac{b_1q^{-1} + b_2q^{-2}}{1 + f_1q^{-1} + f_2q^{-2}} u(t) + e(t) \quad (101)$$

was fitted to the system (100) using a one-step-ahead prediction error method [which in this case is an output error method since $H_*(q) \equiv 1$]. The input was chosen as a PRBS with basic period one sample, which gives $\Phi_u(\omega) \approx 1$ for all ω. Moreover, the filter $L(q)$ was chosen as unity. The limiting estimate is thus given by (71) with $\Phi_u/|H_*|^2 = 1$. The result is shown in Fig. 1.

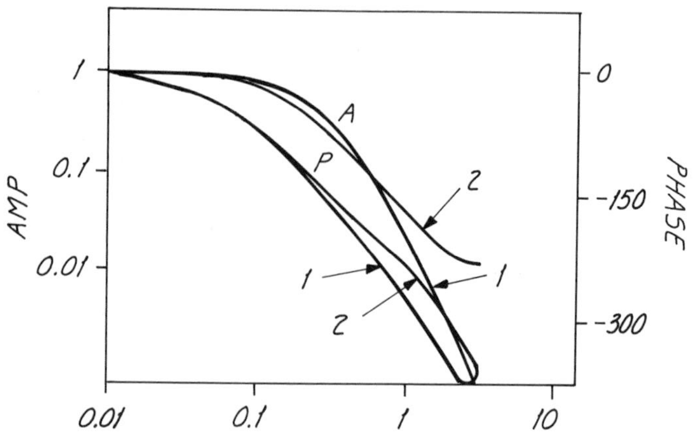

Fig. 1. Bode plots of the true transfer function (100) (curve 1) and the output error model with $\Phi_u/|H_*|^2 = 1$ (curve 2). A, Amplitude; P, phase.

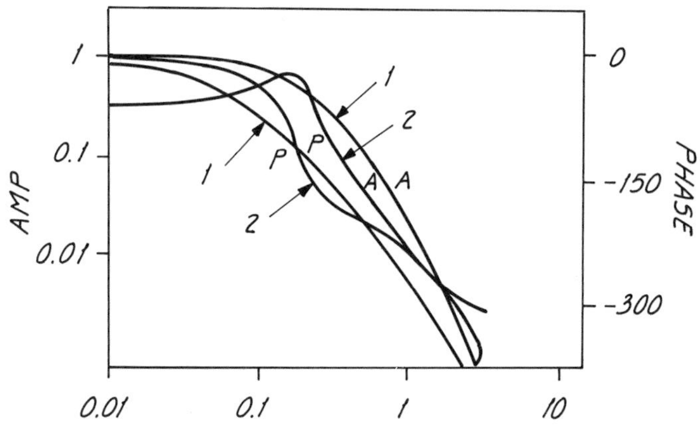

Fig. 2. As Fig. 1, but $\Phi_u(\omega)/|H_*(e^{i\omega})|^2$ *is a high-pass filter.*

To see the effect of prefiltering (or change of noise model), we also applied the same procedure with a prefilter L(q) in (55) (i.e., a noise model $H_* = 1/L$) being a fifth-order high-pass (HP) Butterworth filter with cutoff frequency $\omega_c = 0.1$ rad/sec. This gives the result of Fig. 2. We see that the high-frequency fit is now better at the expense of worse bias at low frequencies.

Example 5.2 Consider the same system as in Example 5.1 with the same input, but let the model set be

$$y(t) = \frac{b_1 q^{-1} + b_2 q^{-2}}{1 + a_1 q^{-1} + a_2 q^{-2}} u(t) + \frac{1}{1 + a_1 q^{-1} + a_2 q^{-2}} e(t). \tag{102}$$

The advantage with this model set is, of course, that the one-step-ahead prediction error method is the simple least squares (LS) method. We will write

$$A(q, \theta) = 1 + a_1 q^{-1} + a_2 q^{-2}$$

and analogously $A_0(q)$ for the true system denominator in (100).

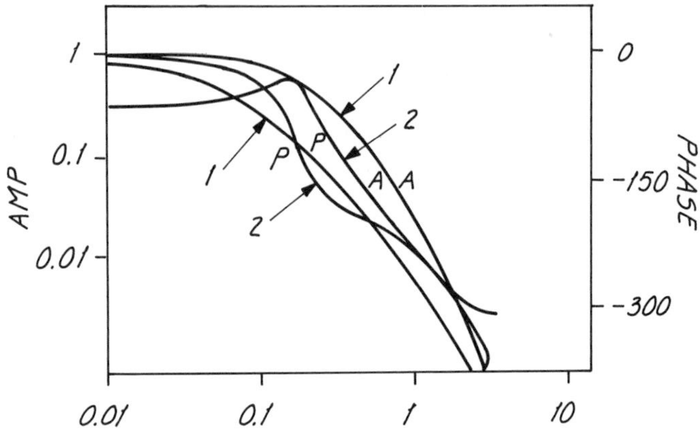

Fig. 3. Bode plots for the true system (curve 1) and the LS estimate using L(q) ≡ 1 (curve 2).

Suppose that the aim of the identification procedure is to obtain good knowledge of the transfer function well around the phase crossover frequency, which for this system will be around 0.2 rad/sec.

Applying the LS method with prefilter $L(q) \equiv 1$ gies the result shown in Fig. 3.

The weighting function in (68), (70) is in this case

$$Q(\omega, \theta^*) = |A(e^{i\omega}, \theta)|^2 \Phi_u(\omega). \tag{103}$$

Since $|A_0(e^{i\omega})|^2$ is a high-pass filter, the high-frequency range is much more important than the low-frequency one in (68). This explains the relatively good fit at high frequencies and the poor one at low frequencies in Fig. 3.

In order to stress that our interest lies in lower frequencies (around 0.2 rad/sec), let us use a filter L in (55) a fifth-order low-pass (LP) Butterworth filter with cutoff frequency ω_c = 0.1 rad/sec, and then make a new LS estimation based upon the filtered data. We then get the Bode plot of Fig. 4.

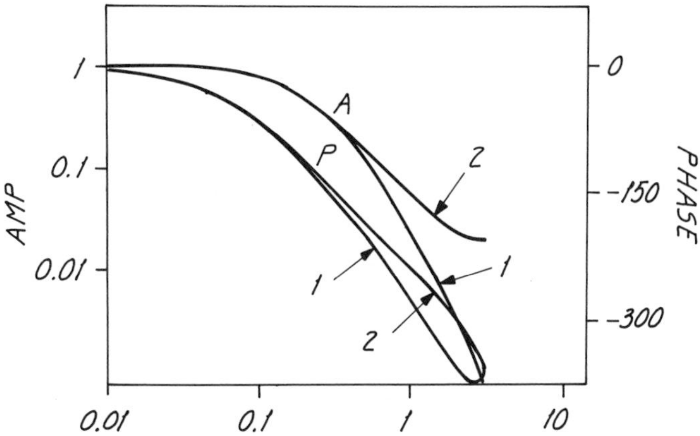

Fig. 4. As Fig. 3, but LP filter applied before the LS identification.

We can see that the low-frequency behavior is much better now. This can be explained by the fact that the weighting function Q in (68) now is

$$Q(\omega, \theta^*) = |L(e^{i\omega})|^2 \Phi_u(\omega)/|A(e^{i\omega}, \theta^*)|^2,$$

which no longer has is of a high-pass character due to the low-pass filter L.

VI. MINIMIZING THE VARIANCE CONTRIBUTION

A. THE PROBLEM

Let us now turn to the problem

$$\min_{\mathcal{D}\in\Delta} J_P(\mathcal{D}), \tag{104}$$

where $J_P(\mathcal{D})$ is given by (81). We shall generally assume that the input power is constrained:

$$\Delta: \int_{-\pi}^{\pi} \Phi_u(\omega)\, d\omega \leq \beta. \tag{105}$$

Spelling out (81) gives

$$J_p(\mathscr{D}) = \int_{-\pi}^{\pi} \Psi(\omega, \mathscr{D})\ d\omega,$$

where

$$\Psi(\omega, \mathscr{D}) = \frac{\lambda C_{11}(\omega) - 2\mathrm{Re}[C_{12}(\omega)\Phi_{eu}(\omega)] + C_{22}(\omega)\Phi_u(\omega)}{\lambda\Phi_u(\omega) - |\Phi_{ue}(\omega)|^2} \cdot \Phi_v(\omega), \tag{106}$$

$$\mathscr{D} = \{\Phi_u, \Phi_{ue}\}. \tag{107}$$

Here we dispensed with the scaling n/N, which is immaterial for the choice of $\mathscr{D}$. Also, the design variable $\mathscr{H}$ (including L and k) does not affect this asymptotic form of the variance contribution.

With the input generation (5), we can use the expressions (59) to rewrite Ψ as (suppressing arguments)

$$\Psi(\omega, \mathscr{D}) = C_{11} \cdot \frac{|1 + FG_0|^2}{\Phi_w} \Phi_v + \frac{\Phi_v}{\Phi_w} 2\mathrm{Re}\ C_{12}FH_0(1 + \overline{FG_0})]$$
$$+ C_{22}\left[\frac{|FH_0|^2}{\Phi_w} \cdot \Phi_v + \frac{1}{\lambda_0}\Phi_v\right]. \tag{108}$$

(Overbar means complex conjugate.) And

$$\mathscr{D} = \{\Phi_w, F\} \tag{109}$$

Supposing that we want to minimize $J_p(\mathscr{D})$, we obtain the following formal problems:

$$\min_{\Phi_w(\omega), F(e^{i\omega})} \int_{-\pi}^{\pi} \Psi(\omega, \mathscr{D})\ d\omega \tag{110}$$

subject to

$$\int_{-\pi}^{\pi} \Phi_u(\omega)\ d\omega = \int_{-\pi}^{\pi} \left\{ \left|\frac{F_0H_0}{1 + FG_0}\right|^2 \lambda_0 + \left|\frac{1}{1 + FG_0}\right|^2 \Phi_w(\omega) \right\} d\omega \leq C \tag{111}$$

(constrained input variance).

We shall not treat this problem in full generality here, but consider some special cases of interest.

B. CASE 1: $C_{12}(\omega) = C_{21}(\omega) = 0$

Consider the case where the $C(\omega)$ matrix is diagonal. This was the case, for example, in Examples 2.1, 2.4 and Example 2.3 (for $\Phi_{ue}^{*} \equiv 0$).

Inserting $C_{11}(\omega) = C_{21}(\omega) = 0$ into (106) gives

$$\min_{\Phi_u, \Phi_{ue}} \int_{-\pi}^{\pi} \frac{\lambda C_{11}(\omega) + C_{22}(\omega)\Phi_u(\omega)}{\lambda\Phi_u(\omega) - |\Phi_{ue}(\omega)|^2} \Phi_v(\omega)\, d\omega \tag{112}$$

subject to the constraint that

$$\int_{-\pi}^{\pi} \Phi_u(\omega)\, d\omega \leq C. \tag{113}$$

From (112) and the fact that $\Phi_{ue}(\omega)$ does not enter the constraint, it follows that

$$\Phi_{ue}^{opt}(\omega) \equiv 0. \tag{114}$$

When $C_{12}(\omega) = 0$, it is thus optimal to use open loop experiments, and the optimal input is easy to compute.

Lemma 6.1. The solution to (112)-(114)

$$\Phi_u^{opt}(\omega) = \mu \cdot \sqrt{C_{11}(\omega) \cdot \Phi_v(\omega)}, \tag{115}$$

where μ is a constant, adjusted so that

$$\int_{-\pi}^{\pi} \Phi_u^{opt}(\omega)\, d\omega = C. \tag{116}$$

Proof. Introduce the constant

$$\Gamma = \int_{-\pi}^{\pi} \sqrt{C_{11}(\omega)\Phi_v(\omega)}\, d\omega.$$

Then, from Schwarz's inequality, we have

$$\Gamma^2 = \left[\int_{-\pi}^{\pi} \sqrt{C_{11}(\omega)\Phi_v(\omega)}\, d\omega\right]^2$$

$$= \left[\int_{-\pi}^{\pi} \frac{\sqrt{C_{11}(\omega)\Phi_v(\omega)}}{\sqrt{\Phi_u(\omega)}} \cdot \sqrt{\Phi_u(\omega)}\, d\omega\right]^2$$

$$\leq \int_{-\pi}^{\pi} \frac{C_{11}(\omega)\Phi_v(\omega)}{\Phi_u(\omega)}\, d\omega \cdot \int_{-\pi}^{\pi} \Phi_u(\omega)\, d\omega$$

$$\leq C \cdot \int_{-\pi}^{\pi} \frac{C_{11}(\omega)\Phi_v(\omega)}{\Phi_u(\omega)}\, d\omega.$$

Hence

$$\int_{-\pi}^{\pi} C_{11}(\omega)\, \frac{\Phi_v(\omega)}{\Phi_u(\omega)}\, d\omega \geq \frac{\Gamma^2}{C}$$

for all $\Phi_u(\omega)$ subject to (116). Equality is obtained for $\Phi_w^{opt}(\omega)$ in (115), which proves the lemma. ∎

With this result we can solve for the optimal input for some of the specific cases discussed in Section II.

Example 6.1. Suppose that we intend to use the model to design a pole placement regulator (40). The corresponding $C(\omega)$ matrix was computed in (42). Consequently, the optimal, constrained variance input is

$$\Phi_u^{opt}(\omega) = \mu \cdot \frac{|R(e^{i\omega})|\sqrt{\Phi_r(\omega)} \cdot |H_0(e^{i\omega})|}{|G_0(e^{i\omega})||1 + G_0(e^{i\omega})P(e^{i\omega})|}. \tag{117}$$

We see that characteristics of the true system are required in order to compute the optimal input. Even though these may not be known in detail, the expression (117) is still useful. It tells us to spend the input energy where (1) a gain increase is desired:

$$\left|\frac{R(e^{i\omega})}{G_0(e^{i\omega})}\right| \gg 1,$$

(2) the reference input is going to have energy: $\sqrt{\Phi_r(\omega)}$ large, (3) the disturbances are significant: $|H_0(e^{i\omega})|$ large, and (4) the sensitivity reduction due to feedback is poor: $|1 + G_0(e^{i\omega})P(e^{i\omega})|$ small. These points are as such quite natural, but their formalization is useful. ■

Example 6.2: Prediction. Suppose that the model is to be applied for prediction when the system operates in open loop with an input spectrum $\Phi_u^*(\omega)$. This means that the corresponding C matrix is given by (34) with $\Phi_{ue}^*(\omega) \equiv 0$. This in turn implies that the optimal, constrained input variance experiment for this application is open loop with input spectrum

$$\Phi_u^{opt}(\omega) = \mu \cdot \sqrt{\Phi_u^*(\omega)}\ . \quad ■ \tag{118}$$

C. *CASE 2: MINIMUM VARIANCE CONTROL*

The C matrix that corresponds to a generalized minimum variance control application of the model is, according to (38),

$$C(\omega) = \left|\frac{R(e^{i\omega})}{H_0(e^{i\omega})}\right|^2 \begin{bmatrix} |M(e^{i\omega})|^2 & M(e^{i\omega}) \\ M(e^{-i\omega}) & 1 \end{bmatrix}, \tag{119}$$

with

$$M(q) = -\frac{H_0(q) - R(q)}{G_0(q)}.$$

The characteristic property of (119) is that it is a singular matrix, and this leads to quite interesting features of the corresponding optimal design problems. (See also [9].)

We find from (106) that

$$\frac{|H_0(e^{i\omega})|^2 \cdot \Psi(\omega, \mathscr{D})}{\Phi_v(\omega) \cdot |R(e^{i\omega})|^2}$$

$$= \frac{\lambda|M(e^{i\omega})|^2 - 2\mathrm{Re}\left[M(e^{i\omega}) - \Phi_{eu}(\omega)\right] + \Phi_u(\omega)}{\lambda\Phi_u(\omega) - |\Phi_{ue}(\omega)|^2}$$

$$= \frac{1}{\lambda}\,\frac{\lambda\Phi_u(\omega) - |\Phi_{ue}(\omega)|^2 + |\lambda M(e^{i\omega}) - \Phi_{ue}(\omega)|^2}{\lambda\Phi_u(\omega) - |\Phi_{ue}(\omega)|^2}$$

$$= \frac{1}{\lambda}\left[+\frac{|\lambda M(e^{i\omega}) - \Phi_{ue}(\omega)|^2}{\lambda\Phi_u(\omega) - |\Phi_{ue}(\omega)|^2}\right],$$

which is minimized for

$$\Phi_{ue}(\omega) = \lambda M(e^{-i\omega}) \tag{120}$$

regardless of $\Phi_u(\omega)$.

Notice that the cross spectrum (120) is realized by

$$u(t) = M(q)e(t) + \tilde{w}(t) \tag{121}$$

regardless of the deterministic input $\tilde{w}$. This corresponds to the feedback law

$$u(t) = \frac{M(q)}{R(q)}\,y(t) + w(t) \tag{122}$$

for an arbitrary input $w[w = (R/H)\tilde{w}]$.

Notice in particular that as long as (122) is an admissible input for some spectrum $\Phi_w(\omega)$, it gives the global minimum

$$\min_{\text{all}\,\mathscr{D}} \int_{-\pi}^{\pi} \Psi(\omega, \mathscr{D})\, d\omega = \frac{1}{\lambda}\int_{-\pi}^{\pi} \frac{|R(e^{i\omega})|^2 \cdot \Phi_v(\omega)}{|M_0(e^{i\omega})|^2}\, d\omega$$

$$= \int_{-\pi}^{\pi} |R(e^{i\omega})|^2\, d\omega$$

of the variance criterion. Hence (122) minimizes $J_p(\mathscr{D})$ with respect to any constraints.

VII. MINIMIZING THE DESIGN CRITERIA

Let us now turn to the full design criterion (25)-(28) in its pragmatic form (79)-(81). Our partial results on bias and variance minimization then show that in certain cases it is possible to minimize the two contributions simultaneously. Then, of course, the full criterion is also minimized. For the case of Theorem 5.1 we thus have the following result.

Theorem 7.1. Consider the problem to minimize (79)-(81) with respect to

$$\mathcal{D} = \left\{\Phi_u(\omega),\ \Phi_{ue}(\omega),\ L(e^{i\omega}),\ H_*(e^{i\omega})\right\}$$

under the assumptions (96)-(97) and subject to the constraint (113). Then $\mathcal{D}_{opt}$ is given by

$$\Phi_{ue}(\omega) \equiv 0,$$

$$\Phi_u(\omega) = \mu_2\sqrt{C_{11}(\omega)\cdot\Phi_v(\omega)}, \qquad (123)$$

$$\left|\frac{L(e^{i\omega})}{\overline{H}_*(e^{i\omega})}\right|^2 = \mu_1\sqrt{\frac{C_{11}(\omega)}{\Phi_v(\omega)}}.$$

Here μ_1 is a constant, adjusted so that the left-hand side has a Laurent expression that starts with a one, and μ_2 is a constant adjusted so that the input power constraint is met. ■

Note that the freedom in the choice of noise model and prefilter is imaginary, since they always appear in the combination $L(q)/H^*(q)$ in the criterion.

The case where our prime interest is in the transfer function G is probably the most common one, and therefore the optimal design variables offered by Theorem 7.1 should be of interest. The only drawback with this solution may be that the choice of constant noise model may lead to more calculations in the numerical minimization of the prediction error criterion.

REFERENCES

1. K. J. ÅSTRÖM and P. EYKHOFF, *Automatica 13*, 457-476 (1971).

2. G. C. GOODWIN and R. L. PAYNE, "Dynamic Systems Identification: Experiment Design and Data Analysis." Academic Press, New York (1977).
Pre

3. P. EYKHOFF, "System Identification." Wiley, London (1974).

4. P. EYKHOFF (Ed.), "Trends and Progress in System Identification." Pergamon Press (1981).

5. L. LJUNG, *Automatica 21*, No. 4, 1985.

6. L. LJUNG, *IEEE Trans. Autom. Control AC-30*, in press (1985).

7. B. WAHLBERG and L. LJUNG, "Design Variables for Bias Distribution," 23rd IEEE Conference on Decision and Control. Las Vegas, Nevada, December (1984).

8. Z. D. YAN and L. LJUNG, *Automatica 21*, No. 4, 1985.

9. M. GEVERS and L. LJUNG, "Benefits of Feedback in Experiment Design," 7th IFAC Symposium on Identification. York, United Kingdom, July (1985).

10. L. LJUNG and Z. D. YUAN, *IEEE Trans. Autom. Control AC-30*, 1985.

11. L. LJUNG, "System Identification—Theory for the User." In press, Prentice-Hall, Englewood Cliffs, New Jersey, 1987.

12. K. J. ÅSTRÖM, "Introduction to Stochastic Control Theory." Academic Press, New York (1970).

13. L. LJUNG, *IEEE Trans. Autom. Control AC-23*, 770-783 (1978).

14. K. J. ÅSTRÖM, *Automatica 16*, 551-574 (1980).

15. L. LJUNG and P. CAINES, *Stochastics 3*, 29-46 (1979).

Techniques in Dynamics Systems Parameter-Adaptive Control

ROLF ISERMANN

Institut für Regelungstechnik
Technical University Darmstadt
6100 Darmstadt, Federal Republic of Germany

I. INTRODUCTION

Adaptive controllers can be designed with a feedforward adaptation (open-loop adaptation or "gain" scheduling) or with a feedback adaptation (closed-loop adaptation). Within the field of adaptive control with feedback there are two basic directions that were followed in the past decade: the concepts of *model reference adaptive control* (MRAC) and *model identification adaptive control* (MIAC). See, for example, the surveys ström [1] and Isermann [2]. [Note that it is difficult to find generally acceptable expressions for the classification of adaptive controllers. Although the term MRAC is well accepted, there are different names for the other class, for example, self-optimizing adaptive controllers (SOAC) or self-tuning regulators (STR). The use of the term MIAC is a proposal to overcome the problems with the interpretation of SOAC and STR.] Model reference adaptive control tries to reach a control behavior close to an a priori given reference model for a definite input variable (e.g., servo control). Model identification adaptive control is based on the identification of a process model and the design of a controller that may optimize an a priori given control performance criterion. Both concepts can be designed by using either *parametric process models* (e.g., differential or difference equations of finite order) or *nonparametric models* (e.g., impulse responses). Further, the used models and controllers can be formulated for *continuous time* or *discrete time*.

In this article a review is given on our development of MIAC based on parametric process models in discrete time. The resulting control systems are called *parameter-adaptive control systems* (PACs). They are frequently also called self-tuning regulators (STRs). Figure 1 shows the basic block diagram.

One of the earliest approaches to MIAC was described by Kalman [3], who combined the parameter estimation method of least squares with a deadbeat controller. Ten years later the availability of process computers allowed broader development. Parameter-adaptive controls were proposed by Peterka [4] and Åstrom and Wittenmark [5], who used the recursive least squares method together with a minimum variance controller. Clarke and

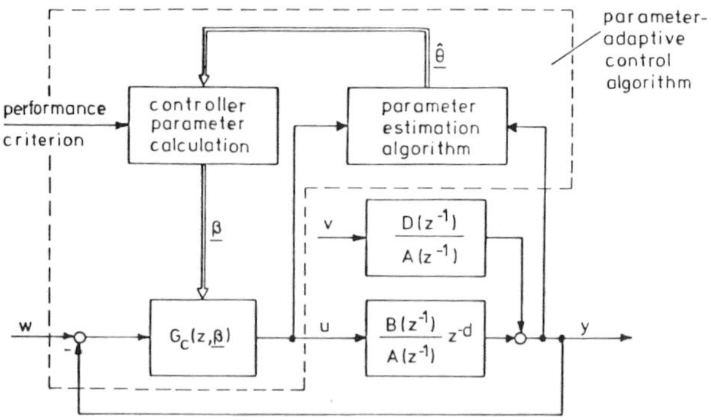

Fig. 1. Basic block diagram for parameter-adaptive controllers.

Gawthrop [6] introduced a modified minimum variance controller, which allows to weight the control input. Kurz *et al.* [7] then investigated explicit combinations of three recursive parameter estimators with four types of control algorithms. Wellstead *et al.* [8] treated recursive least squares with pole/zero assignment controllers. For further references see Åström [1].

II. APPROACHES TO PARAMETER-ADAPTIVE CONTROL

A. *BASIC ELEMENTS*

Figure 1 indicates that the basic elements of a parameter-adaptive controller are (1) the process model, (2) the parameter estimation method, (3) information about the process, (4) criterion for controller design, and (5) control algorithms. There are many possibilities in selecting these elements. In the following discussion the elements that finally turned out to be well suited with respect to theoretical and practical requirements are briefly described.

1. *Process Model*

Parametric process models in the form of finite-order difference equations have several advantages. First, parameter estimation methods can be used, which are powerful in noisy situations and may satisfy the closed-loop identifiability conditions. Second, they enable a simple inclusion of dead times and stochastic noise models. Third, they allow the direct use of modern controller design methods. For linearizable processes, the following stochastic difference equation has shown to be well suited:

$$y(k) = \underline{\psi}^T(k)\ \underline{\theta} + v(k), \tag{1}$$

with the vectors

$$\underline{\psi}^T(k) = [-y(k-1) \cdots -y(k-m) \mid u(k-d-1) \cdots u(k-d-m) \mid v(k-1) \cdots v(k-m)], \tag{2}$$

$$\underline{\theta} = [a_1 \cdots a_m \mid b_1 \cdots b_m \mid d_1 \cdots d_m]^T, \tag{3}$$

where $k = t/T_0 = 0, 1, 2$ is the discrete time, T_0 the sampling time, d the discrete dead time, and

$$y(k) = U(k) - Y_{00}, \quad u(k) = Y(k) - U_{00} \tag{4}$$

are the deviations of the measured process output signal Y(k) and input signal U(k) from the direct current (d.c.) values Y_{00} and U_{00}. The unmeasurable noise v(k) is assumed to be statistically independent and stationary with zero mean. The corresponding z transfer function of Eq. (1) is

$$y(z) = \underbrace{\frac{B(z^{-1})}{A(z^{-1})}}_{G_p(z^{-1})} z^{-d} u(z) + \underbrace{\frac{D(z^{-1})}{A(z^{-1})}}_{G_v(z^{-1})} v(z), \tag{5}$$

with the polynomials

$$\begin{aligned} A(z^{-1}) &= 1 + a_1 z^{-1} + \cdots + a_m z^{-m}, \\ B(z^{-1}) &= b_1 z^{-1} + \cdots + b_m z^{-m}, \\ D(z^{-1}) &= 1 + d_1 z^{-1} + \cdots + d_m z^{-m}, \end{aligned} \tag{6}$$

where $z = \exp(T_0 s)$ with $s = \delta + i\omega$ the Laplace variable. By assuming $D(z^{-1}) = 0$ a deterministic process model results. A further advantage of the process model Eq. (1) or (5) is that it can straightforwardly be extended to multivariable and nonlinear processes (see Section X).

2. Parameter Estimation Methods

For on-line estimation of the unknown process parameters $\underline{\theta}$ in real time, parameter estimation methods are used, for example, recursive least squares (RLS) or recursive extended least squares (RELS) and their square-root filter representations. This is treated in Section III.

3. Information about the Process

If the estimated process parameters $\hat{\underline{\theta}}$ are assumed to be identical with the real process parameters $\underline{\theta}$, the resulting controllers are called certainty equivalence controllers. This approach is rather simple and has proved to be sufficient in many cases. As alternative, the uncertainties $\Delta\underline{\theta}$ can be taken into account.

4. Criterion for Controller Design

Only nondual adaptive controllers are considered here, because just the present and past information about the process is used for controller design. Linear quadratic performance criteria are very well suited from the viewpoints of design and practicability. Pole placement techniques are an alternative if it is known where to place the poles (servo systems, vehicles). Other criteria such as finite settling time or pole-zero cancellation may be used in special cases.

5. Control Algorithms

Control algorithms for adaptive control should satisfy the closed-loop identifiability conditions and should require a small computational effort for the controller parameter calculation. This means that the order of the control algorithm depends on the process model order and that the actual design of the control algorithm is done before the implementation. Then, for example, state controllers, deadbeat controllers, or minimum variance controllers are suited. However, the classical PID controller for model orders $m > 2$ does not satisfy these requirements. Despite this, it is also possible to design parameter-adaptive PID controllers. This subject is treated in Section IV.

B. *SUPPLEMENTARY ELEMENTS*

The development of parameter-adaptive control algorithms has shown that it is not sufficient just to add the basic elements together. For example, the following items have to be taken into account: estimation of signal d.c. values, compensation of offsets for control algorithms without integral action, numerical properties and improvements, supervisory functions, and aids for the specification of design parameters. This will be included in the sequel.

C. *PRACTICAL REQUIREMENTS*

There are some other conditions that should be satisfied to develop adaptive controllers and to apply them.

1. *Modularity*

The discussion has already shown that depending on the process, the noise, the required control performance, and so on, different elements have to be combined properly. Therefore it is advisable to program modules for these elements in order to reach an easy configuration depending on the needs and to make possible later improvements easier.

2. *Extensibility*

The concept of parameter-adaptive control is not limited to linear single-input/single-output processes. For example, the same principle can be used for multivariable processes or for the nonlinear control of nonlinear processes. Therefore the extensibility of the used models, parameter estimators, and controllers to these further tasks is another consideration.

3. Robustness

For each application the robustness of the control performance including all aspects such as proper selection of parameter estimators and controllers, numerical properties, noise reduction, and selection of starting parameters is of importance. The robustness should, of course, be better than with classical fixed controllers.

4. Access to Intermediate Results

At least for the commissioning phase and for continuously acting supervisory functions, the access to intermediate results can be meaningful. For example, the process parameters or the process transient function or the process poles and zeros, the covariance matrix of the parameter estimation, and the equation error or its correlation function give an indication of the inner state.

5. Acceptability

In order to introduce adaptive control into practice several additional functions should be offered by the equipment. This has mainly to do with the understanding, the relation to commonly used controllers, and operational instructions. Therefore the design of the man-machine interface is important.

III. PARAMETER ESTIMATION

A. *DYNAMIC PARAMETERS*

If the d. c. values U_{00} and Y_{00} are assumed to be known, the dynamic model parameters $\hat{\underline{\theta}}$ can be estimated by the nonrecursive least squares (LS) method:

$$\hat{\underline{\theta}} = [\underline{\Psi}^T\underline{\Psi}]^{-1}\underline{\Psi}^T\underline{y} = \underline{P} \cdot \underline{\Psi}^T\underline{y}, \tag{7}$$

where $\underline{P} = [\underline{\Psi}^T\underline{\Psi}]^{-1}$ is proportional to the covariance matrix of the parameter estimates and $\underline{\Psi}^T$ and $\underline{y}$ contain the measured signals (see, e.g., Eykhoff [9]). For on-line identification in real time, a convenient way is to bring Eq. (7) in a recursive form. Then the recursive least squares method results. With recursive estimation a noise filter model can also be included, which leads to the recursive extended least squares method. These recursive estimation algorithms and others can be written in a unified form, as in Söderström *et al.* [10], Isermann [11], Strejc [12]:

$$\hat{\underline{\theta}}(k + 1) = \hat{\underline{\theta}}(k) + \underline{\gamma}(k)e(k + 1), \tag{8}$$

$$\underline{\gamma}(k) = \mu(k + 1)\underline{P}(k)\underline{\varphi}(k + 1), \tag{9}$$

$$e(k + 1) = y(k + 1) - \underline{\psi}^T(k + 1)\hat{\underline{\theta}}(k). \tag{10}$$

The definitions of $\hat{\underline{\theta}}$, $\underline{\psi}^T$, $\underline{\varphi}$, and $\underline{P}$ depend on the parameter estimation method. For RLS, it is ($d_i = 0$, $v(k - i) = 0$, $i = 1, 2, \ldots, m$):

$$\underline{\varphi}(k + 1) = \underline{\psi}(k + 1), \tag{11}$$

$$\mu(k + 1) = \left[\lambda(k + 1) + \underline{\psi}^T(k + 1)\underline{P}(k)\underline{\psi}(k + 1)\right]^{-1}, \tag{12}$$

$$\underline{P}(k + 1) = \left[\underline{I} - \underline{\gamma}(k)\underline{\psi}^T(k + 1)\right]\underline{P}(k)/\lambda(k + 1), \tag{13}$$

with $\lambda(k)$ a forgetting factor, $0 < \lambda(k) < 1$, if slowly varying process parameters have to be tracked. If all convergence conditions of RLS are satisfied, $\underline{P}(k)$ is a standardized estimate of the covariance matrix of the parameter estimate errors

$$E\{\underline{P}(k)\} = \mathrm{cov}[\Delta\underline{\theta}(k - 1)]/\sigma_e^2, \tag{14}$$

with $\Delta\underline{\theta} = \hat{\underline{\theta}} - \underline{\theta}_0$ ($\underline{\theta}_0$ true parameters) and σ_e^2 the variance of the equation error $e(k)$, Eq. (9).

If RELS is used, the parameters d_i of the noise model are also obtained. However, they converge slower than the process model parameters. Therefore RELS should only be used for large and stationary noise $n(k)$.

With respect to parameter-adaptive control, several modifications of the basic parameter estimation algorithms were developed in order to improve the numerical properties for finite-word-length microcomputers, the access to intermediate results, and the influence of assumed starting values. Of special importance are the use of "square-root filter" implementations (see, e.g., Biermann [13], Strejc [12], Radke [14]). There are mainly two different ways to introduce square-root filtering. They are based either on the covariance or on the information matrix representation of the parameter estimation algorithms. The symmetric covariance matrix $\underline{P}$ can be separated in two triangular matrices

$$\underline{P} = \underline{S}\underline{S}^T, \tag{15}$$

where $\underline{S}$ is called the square root of $\underline{P}$. The resulting *discrete square-root filter algorithms in the covariance form* (DSFC) are

$$\hat{\underline{\theta}}(k + 1) = \hat{\underline{\theta}}(k) + \underline{\gamma}(k)e(k + 1),$$

$$\underline{\gamma}(k) = a(k)\underline{S}(k)\underline{f}(k),$$

$$
\begin{aligned}
\underline{f}(k) &= \underline{S}^T(k)\underline{\psi}(k+1), \\
\underline{S}(k+1) &= \left[\underline{S}(k) - g(k)\underline{\gamma}(k)f^T(k)\right]\frac{1}{\sqrt{\lambda(k)}}, \\
\frac{1}{a(k)} &= \underline{f}^T(k)\underline{f}(k) + \lambda(k), \\
g(k) &= \frac{1}{[1 + \sqrt{\lambda(k)a(k)}]},
\end{aligned} \tag{16}
$$

with starting value $\underline{S}(0) = \sqrt{\alpha} \cdot \underline{I}$. These equations appear in similar form for state estimation in Kaminski *et al.* [15].

The *discrete square-root filter algorithm in the information form* (DSFI) follows from the nonrecursive LS by writing

$$\underline{P}^{-1}(k+1)\hat{\underline{\theta}}(k+1) = \underline{\Psi}^T(k+1)\underline{y}(k+1) = \underline{f}(k+1), \tag{17}$$

where the information matrix originally is updated recursively by

$$
\begin{aligned}
\underline{P}^{-1}(k+1) &= \lambda(k+1)\underline{P}^{-1}(k) + \underline{\psi}(k+1)\underline{\psi}^T(k+1), \\
\underline{f}(k+1) &= \lambda(k+1)f(k) + \underline{\psi}(k+1)y(k+1).
\end{aligned}
$$

To avoid possible numerical problems through ill conditioning and the matrix inversion in Eq. (17), $\underline{P}^{-1}(k+1)$ should be presented in a triangular form

$$\underline{P}^{-1} = (\underline{S}^{-1})^T\underline{S}^{-1}, \tag{18}$$

so that $\hat{\underline{\theta}}(k+1)$ follows directly from

$$\underline{S}^{-1}(k+1)\hat{\underline{\theta}}(k+1) = \underline{b}(k+1) \tag{19}$$

by backward processing. This is obtained by an orthogonal transformation matrix $\underline{T}$ (with $\underline{T}^T\underline{T} = \underline{I}$) applied to Eq. (7),

$$\underline{\Psi}^T\underline{T}^T\underline{T\Psi\hat{\theta}} = \underline{\Psi}^T\underline{T}^T\underline{Ty} \tag{20}$$

where

$$\underline{T\Psi} = \begin{bmatrix} \underline{S}^{-1} \\ \underline{0} \end{bmatrix} \tag{21}$$

has an upper triangular form and

$$\underline{T}\underline{y} = \begin{bmatrix} \underline{b} \\ \underline{w} \end{bmatrix}. \tag{22}$$

From Eq. (20) follows

$$\underline{T}(k+1)\underline{\Psi}(k+1)\underline{\hat{\theta}}(k+1) = \underline{T}(k+1)\underline{y}(k+1). \tag{23}$$

This turned into a recursive form results in [see Kaminski and others (1971)]

$$\begin{bmatrix} \underline{S}^{-1}(k+1) \\ \underline{0}^T \end{bmatrix} = \underline{T}(k+1)\begin{bmatrix} \sqrt{\lambda}\underline{S}^{-1}(k) \\ \underline{\psi}^T(k+1) \end{bmatrix}, \tag{24}$$

$$\begin{bmatrix} \underline{b}(k+1) \\ w(k+1) \end{bmatrix} = \underline{T}(k+1)\begin{bmatrix} \sqrt{\lambda}\underline{b}(k) \\ y(k+1) \end{bmatrix}. \tag{25}$$

Then $\underline{S}^{-1}(k+1)$ and $\underline{b}(k+1)$ are used to calculate $\underline{\hat{\theta}}(k+1)$ via Eq. (19). Note that no starting value $\underline{\hat{\theta}}(0)$ has to be assumed and therefore the initial convergence is very good. Further, no matrix inversion is required. Several simulations and practical applications have shown that the general properties (numerics, convergence) of DSFI are better than those of DSFC or RLS (Radke [14]).

B. STATIC PARAMETERS

To tackle the problem with the normally unknown d.c. values, Eq. (4), two ways have shown to be useful for PACs.

1. Implicit Method

Introducing Eq. (4) in Eq. (1) leads to the process model $[v(k) = 0]$,

$$Y(k) = \underline{\psi}_*^T(k)\underline{\theta}_*, \tag{26}$$

with

$$\underline{\psi}_*^T(k) = [-Y(k-1) \cdots -Y(k-m) \mid U(k-d-1)$$

$$\cdots U(k-d-m) \mid 1], \tag{27}$$

$$\underline{\theta}_* = \left[\hat{a}_1 \cdots \hat{a}_m \mid \hat{b}_1 \cdots \hat{b}_m \mid \hat{C}\right]^T \tag{28}$$

$$C = (1 + a_1 + \cdots + a_m)Y_{00} - (b_1 + \cdots + b_m)U_{00}. \tag{29}$$

Hence by introducing the constant C, the absolute signal values U(k) and Y(k) can be used directly. The knowledge of $\hat{C}$ can be applied to compensate offsets of controllers with no integral action. If it is assumed that $Y_{00} = W(k)$ (W, set point) for proportional action processes, the required U_{00} can be calculated by Eq. (29) to remove the offset. However, this depends on the accuracy of all parameter estimates. A disadvantage of the implicit method is that dynamic parameters $\hat{\underline{\theta}}$ change if the static parameter $\hat{C}$ changes, and vice versa.

2. *Explicit Method*

By taking the first differences

$$\Delta y(k) = Y(k) - Y(k-1); \quad \Delta u(k) = U(k) - U(k-1), \tag{30}$$

the d.c. values disappear. The dynamic parameters $\underline{\theta}$ then can be obtained by using

$$\underline{\psi}^T = [-\Delta y(k-1) \cdots -\Delta y(k-m) \mid \Delta u(k-d-1)$$

$$\cdots \Delta u(k-d-m)]. \tag{31}$$

If the constant C has to be known, a least squares estimate is

$$C(k) = \frac{1}{k+1} \sum_{i=0}^{k} L(i), \tag{32}$$

$$L(i) = Y(i) + \hat{a}_1 Y(i-1) + \cdots - \hat{b}_1 U(i-d-1)$$

$$- \cdots -b_m U(i-d-m). \tag{33}$$

In this case the dynamic parameters $\hat{\underline{\theta}}$ are independent on the estimate C, but the procedure is sensitive to high-frequency noise. Therefore proper filtering has to be used beforehand.

IV. CONTROL ALGORITHM DESIGN

A general linear controller has the transfer function

$$G_R(z) = \frac{u(z)}{e_w(z)} = \frac{Q(z^{-1})}{P(z^{-1})} = \frac{q_0 + q_1 z^{-1} + \cdots + q_\nu z^{-\nu}}{p_0 + p_1 z^{-1} + \cdots + p_\mu z^{-\mu}}, \tag{34}$$

with the control deviation

$$e_w(k) = w(k) - y(k) \tag{35}$$

and w(k) a variation of the reference value W(k). If the only exciting input signal for the closed loop is the unmeasurable stochastic noise v(k), the process parameters only then converge to unique values if a closed-loop identifiability condition is satisfied

$$\max\{\mu;\ \nu + d\} - p \geq m, \tag{36}$$

where p is the number of common poles and zeros in the closed-loop transfer function. This means that the controller order must be sufficiently large (see, e.g., Gustavsson *et al.* [16]). Additionally, the computational effort for controller parameter calculation should be small. These properties are met by the following controllers.

A. *DEADBEAT CONTROLLER*

The transfer function

$$G_{DB}(z) = \frac{q_0 A(z^{-1})}{1 - q_0 B(z^{-1}) z^{-d}}, \quad q_0 = \left[\sum_{i=1}^{m} b_i\right]^{-1} \tag{37}$$

shows that its parameters are calculated with a very small effort. However, it should only be used with increased order for stable low-pass processes (see [11]).

B. MINIMUM VARIANCE CONTROLLER

The design equations for minimizing the performance function

$$\underline{I}(k + d + 1) = E\{y^2(k + d + 1) + ru^2(k)\}$$

are

$$G_{MV}(z) = -\frac{L(z^{-1})}{zB(z^{-1})F(z^{-1}) + D(z^{-1})r/b_1}, \tag{38}$$

$$D(z^{-1}) = A(z^{-1})F(z^{-1}) + z^{-(d+1}L(z^{-1}).$$

This controller may be used for a larger class of processes if stochastic noise acts on the process.

C. STATE CONTROLLERS

If the process is described by the state-variable model

$$\underline{x}(k + 1) = \underline{Ax}(k) + \underline{b}u(k), \quad y(k) = c^T\underline{x}(k),$$

the basic state controller is

$$u(k) = -\underline{k}^T\hat{\underline{x}}(k), \tag{39}$$

where the state variables are usually determined by an observer. If the state-variable model is written in the row-companion canonical form, the state variables can be directly calculated from the output signal y(k) (Schumann [17], [18]). The calculation of the gain vector may be performed by a recursive solution of the matrix Riccati equation. Usually about 10 steps are sufficient to make the computational effort is acceptable (Radke [14]). There are some other details that have to be implemented to reach a good state control. (See, e.g., [11].)

From a theoretical standpoint deadbeat (DB) and minimum variance (MV) controllers and state controllers (SCs) meet the basic conditions. However, because process engineers and the operating personnel are not familiarized with these more complex controllers, acceptance problems may arise. Adaptive PID controllers would be much more welcome. But PID controllers theoretically do not satisfy the closed-loop identifiability for $m > 2 + d$ and the controller parameter calculation is not straightforward for $m > 2$.

D PID CONTROLLERS

The simplest transfer function of a PID control algorithm is

$$G_{PID}(z) = \frac{q_0 + q_1 z^{-1} + q_2 z^{-2}}{1 - z^{-1}}. \tag{40}$$

In Radke and Isermann [19] various design methods are discussed. Pole assignment, cancellation principle, and approximation of other controllers are restricted to special cases. A more general approach is to minimize a quadratic performance criterion by a parameter optimization method. To save calculations, the performance criterion is calculated by recursive formula in the z domain. The Hooke-Jeeves method is used for hill climbing and the calculations are distributed over several sampling intervals. Simulations and applications have shown that the convergence is comparable to the other PACs and that the violation of the closed-loop identifiability condition in practice did not give problems. (For more details see [19].) A simpler approach based on tuning rules is described in Kofahl and Isermann [20].

V. COMBINATIONS

As discussed in Section II, parameter-adaptive controllers result from proper combinations of dynamic parameter estimators, static parameter estimators, control algorithms, and offset compensators. There are different ways to organize the combinations of recursive parameter estimation algorithms and control algorithms and the synchronization of both.

A. *EXPLICIT AND IMPLICIT COMBINATIONS*

The process model be described by

$$y(k) = \underline{\psi}^T(k)\underline{\theta}(k - 1) \tag{41}$$

and the control algorithm by

$$u(k) = \underline{\rho}^T(k)\underline{\Xi}(k - 1). \tag{42}$$

In the case of an *explicit combination*, the process parameters $\hat{\underline{\theta}}$ are estimated explicitly and stored as an intermediate result. Then the controller parameters are calculated

$$\underline{\Xi}(k) = f\left[\hat{\underline{\theta}}(k)\right], \tag{43}$$

and the new process input $u(k + 1)$ follows from Eq. (42). For an *implicit combination* the controller design equation (43) is introduced into the process model (41), so that

$$y(k) = \underline{\zeta}^T(k)\underline{\Xi}(k - 1). \tag{44}$$

Then the controller parameters $\underline{\Xi}(k)$ are estimated directly by a recursive algorithm. In this case the process parameters are contained implicitly and do not appear as intermediate results (see also [1], [2]).

The advantages of the *implicit combination* are that by circumventing the controller design equations some calculation time may be saved and that the theories for the convergence of recursive parameter estimation can be directly applied. However, the number of parameters to be estimated may increase, and only certain estimators and controllers can be written in an implicit scheme. The first proposals of Peterka [4], Åström and Wittenmark [5], and Clarke and Gawthrop [6] were implicit combinations.

The *explicit combination* offers basically more freedom for the design. It allows many different combinations (see Table I and Kurz *et al.* [7]), enables modular programming, and allows direct access to the process parameter estimates and performance measures of their estimators, which is important for the required supervisory functions and for the acceptability. Also extensions to multivariable and nonlinear processes are directly possible and later modifications in the basic elements are much easier to perform. Because of these more flexible and more transparent properties, we prefer the explicit combination.

B. *SYNCHRONOUS AND ASYNCHRONOUS COMBINATION*

The original versions of (explicit) PAC use the following equations sequentially after measuring the new control variable y(k). For the control algorithm,

$$u(k) = \underline{\rho}^T(k)\underline{\Xi}(k-1), \tag{45}$$

for the parameter estimator,

$$\underline{\hat{\theta}}(k) = \underline{\hat{\theta}}(k-1) + \underline{\gamma}(k-1)e(k), \tag{46a}$$

and for the controller parameter,

$$\underline{\Xi}(k) = f[\underline{\hat{\theta}}(k)], \tag{46b}$$

TABLE I. *Applicability of Different Parameter-Adaptive Controllers*

Parameter estimator/controller	Type of process			Type of disturbance	
	Asymptotic stable	Integral behavior	Zeros outside unit circle	Mainly stochastic $n(k)$	Mainly deterministic $w(k)$
RLS-DB[a]	x	--	x	--	x
RLS-MV ($r = 0$)[a]	x	x	--	x	--
RLS-MV ($r > 0$)[a]	x	x	x	x	x
RLS-PID	x	x	x	x	x
RLS-SC	x	x	x	x	x
RELS-MV ($r > 0$)	x	x	x	x	x

[a] Explicit and implicit combinations known.

with (46a) and (46b) for the next step [to calculate u(k + 1)]. Hence parameter estimation, controller parameter calculation, and the control algorithm have the same sampling time and are performed within one sampling interval. This may be called a *synchronous combination.*

There are now several other ways to design the interface between the parameter estimator and the controller, which result in *asynchronous combinations.*

1. Different Sampling Times

The control algorithm and parameter estimator work with different sampling times. For example, the control algorithm has a small sampling time T_{0c} in order to reach a good control performance and the parameter estimator a larger one ($T_{0p} = \kappa T_{0c}$, $\kappa = 2, 3, \ldots$) because of better numerical properties. For fast processes two different microprocessors may be used for control and parameter estimation. Or the controller design may be distributed over several sampling intervals if there are time problems. Also fast sampling for parameter estimation and slow sampling for control can be arranged. This was, for example, used to search on line an appropriate sampling time for the controller (Schumann *et al.* [21]).

2. Conditional Controller Design

The calculation or the change of the controller parameters may be performed only if certain conditions are fulfilled. For example, if the process model parameters have exceeded a certain threshold or a persistently exciting process input is acting or the closed-loop simulation gives a better control performance than the present one.

Hence, there are many possibilities to design the way of combining the parameter estimator and the controller, depending on the process, the signals, the actual need for adaptation, and the computational capacity.

VI. CONVERGENCE CONDITIONS

A survey on the stability and convergence of adaptive controllers is given in Åström [1] as far as general results are known hitherto (see also [2]). Therefore only some convergence conditions are discussed here, which are important for the design and the supervision of PAC.

A first necessary condition is that the real process with the fixed controller is stabilizable with the exact controller parameters and their values in the neighborhood. The convergence analysis of adaptive control systems may be divided into three steps [22]: (1) the convergence at the beginning, (2) the convergence far from the convergence point, and (3) the convergence near the convergence point (asymptotic convergence). In all cases the convergence rates and the convergence points are of interest.

Sufficient conditions for the asymptotic convergence of explicit PAC follow from the requirement that the process parameters $\hat{\underline{\theta}}$ converge to such values $\underline{\theta}_{ss}$ that the controller parameters $\underline{\Gamma}$ converge to the exactly tuned values $\underline{\Gamma}_0$, that is,

$$\lim_{k\to\infty} E\{\underline{\Gamma}(k)\} = \underline{\Gamma}_0. \tag{47}$$

This is in general the case if the process model parameters converge to the true values, $\underline{\theta}_{ss} = \underline{\theta}_0$. Hence, consistent parameter estimates

$$\lim_{k\to\infty} E\left\{\hat{\underline{\theta}}(k)\right\} = \underline{\theta}_0 \tag{48}$$

in mean square are required. (Exceptions are discussed later.) This is surely the case if the following necessary conditions are satisfied.

(1) The process is stable and identifiable and can be described accurately enough by a linear difference equation with constant parameters.

(2) The process order m and dead time d are exactly known.

(3) The conditions for closed-loop identifiability are satisfied.

(4) $e(k)$ is uncorrelated with $u(k)$.

(5) $e(k)$ is uncorrelated and $E\{e(k)\} = 0$.

(6) $\lim_{k\to\infty} \frac{1}{k} \underline{P}^{-1}(k) = \lim_{k\to\infty} \underline{H}(k)$

is positive definite [which implies a persistently exciting process input $u(k)$ of order m].

(7) For RELS, $H(z) = 1/D(z^{-1}) - \frac{1}{2}$ is positive real.

Depending on the parameter-adaptive controller being considered, some of these conditions may be weakened or even changed. For example, (1) the parameter estimates must not necessarily converge to the true values [$e(k)$ is correlated]. Biased parameters may be tolerated (or even lead to asymptotic convergence, e.g., RLS/MV). (2) The process may be unstable. (3) The conditions for closed-loop identifiability may be circumvented by assuming certain controller parameters to be known (MV controllers).

The conditions also depend on the type of disturbances. If stochastic disturbances n(k) act on the process, the process input u(k) must be persistently exciting of order m. However, for initial deviations from steady-state and missing disturbances, this condition may not be necessary.

For some parameter-adaptive controllers it was possible to give some more rigorous convergence proofs. Asymptotic convergence for the implicit RLS-MV was demonstrated by Åström and Wittenmark [5] and Ljung [23]. Matko and Schumann [24] have shown a global convergence for the explicit RLS-DB in the case of a deterministic reference input. Schumann [18] has given convergence conditions far from the convergence point for explicit PAC based on RLS for deterministic and RELS for stochastic disturbances.

If the process parameters are slowly time varying, the parameter-adaptive control algorithms can be applied, too, if a persistently exciting external signal acts on the loop and a forgetting factor $\lambda < 1$ is chosen properly.

VII. STEPS FOR THE APPLICATION

A. CHOICE OF THE PARAMETER-ADAPTIVE ALGORITHM

Parameter-adaptive control algorithms may be applied for (1) self-tuning of controllers during the implementation phase and (2) adaptive control of slowly time-varying processes. In case 1 it is possible to tune controllers, which are later on fixed, in short time with rather high accuracy. Then especially RLS-PID and RLS-SC are of interest because the loops with other controllers are more sensitive to process parameter changes after the implementation phase. For this self-tuning in a limited time period, the closed-loop identifiability condition (36) must not be satisfied. Usually a forgetting factor $\lambda = 1$ should be taken.

For slowly time-varying processes or for weak nonlinear processes that can be linearized around slowly changing operating points, the parameter-adaptive control algorithms of Table I can be applied as adaptive controllers (case 2) if a forgetting factor $\lambda < 1$ is chosen.

By choosing an appropriate λ one has to compromise between the abilities of the parameter estimator to reduce the effect of the process noise and to follow time-varying parameters. The choice of the parameter-adaptive control algorithms in dependence on the process and the type of disturbances follows from Table I. However, the applied algorithm may also depend on the required calculation time in comparison to the permitted sampling time T_0, which depends on the settling time of the process.

B. PREIDENTIFICATION

Because parameter-adaptive control is based on process parameter estimation, it must be ensured that the behavior of the process, the applied parameter estimation method, and the various factors enable the determination of an accurate process model. Therefore it is recommended that in the case of an unknown process, first an identification experiment is performed, open loop for stable processes and closed loop with a fixed controller for unstable processes. A perturbation signal is applied to the process input, and after a sufficiently long identification time, the obtained model has to be verified; that is, it has to be checked if the identified model agrees with the real process. This may also include the search for the sampling time T_0, the model order $\hat{m}$, and dead time $\hat{d}$. For a summary of various ways of model verification see, for example, [25]. Because identification is generally an iterative procedure, this holds also, at least for the starting phase of a parameter-adaptive controller.

C. *CHOICE OF DESIGN FACTORS*

To start the adaptive control algorithms the following factors have to be specified a priori: T_0, sampling time; $\hat{m}$, process model order; $\hat{d}$, process model dead time; λ, forgetting factor; and r, process input weighting factor.

In general, digital parameter-adaptive control is not very sensitive to the choice of the sampling time T_0. For proportional action processes good control can be obtained mostly within the range

$$T_{95}/20 \leqq T_0 \leqq T_{95}/5, \tag{49}$$

where T_{95} is the 95% settling time of the process transient response. To avoid process input changes that are too strong, the sample time should not be chosen too small for DB and MV controllers.

Simulations with a process of order $m_0 = 3$ have shown that the adaptive control was not sensitive to wrong model orders with in the range

$$m_0 = 1 \leqq \hat{m} \leqq m_0 + 2. \tag{50}$$

Also other simulations have shown that the order needs not be known exactly.

However, the adaptive control algorithms are sensitive to the choice of the dead time $\hat{d}$. If d is not known exactly or changes with time, the control is either poor or unstable, especially for combinations with minimum variance controllers. But this can be overcome by including a dead-time estimation, as shown by Kurz [26].

The choice of the forgetting factor λ for the parameter estimation depends on the speed of process parameter changes, the model order, and the kind of disturbances. For constant processes or very slowly time-varying processes, $\lambda = 0.99$ is recommended. For slowly time-varying processes and stochastic disturbances, $0.95 \leqq \lambda \leqq 0.99$ and stepwise reference variable changes $0.85 \leqq \lambda \leqq 0.90$ are proper choices.

The influence of the weighting factor r on the manipulated variable can be estimated as described in [2]. This factor can also be tuned manually during the implementation phase.

D. *STARTING PROCEDURES*

The parameter-adaptive controller may be switched on from the very beginning. However, then during the first adaptation phase an unpredictable control can be obtained, depending on the initial state and disturbances (Fig. 2a). To limit the changes of the process input upper and lower restrictions (by software) can be set. Another way is to introduce a perturbation signal to the process input for a short period (10 to 20 samples) in order to obtain a reasonable starting model, before closing the loop with the control algorithm, as described for the preidentification. This leads to a well-defined starting phase (Fig. 2b) with limited oscillations of u(k) and y(k).

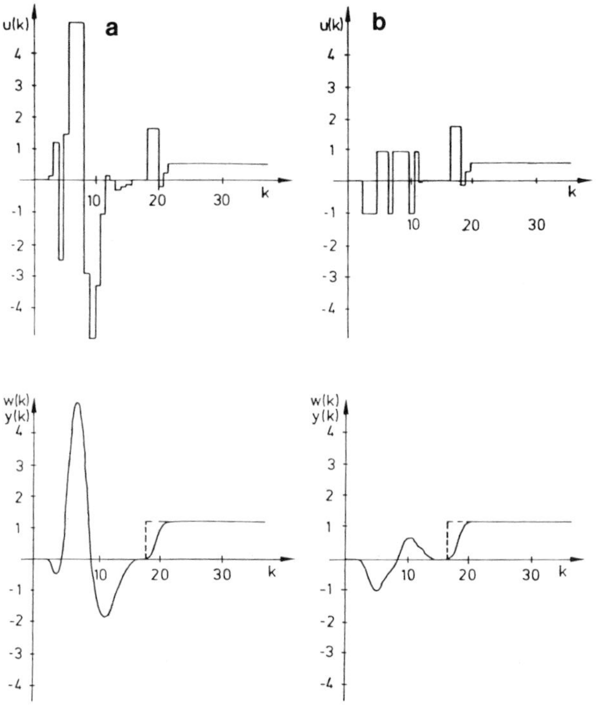

Fig. 2. Starting procedures for RLS-DB with m = 3; d = 0; T_0 = 2 sec; λ = 0.95. (a) Without perturbation and closed loop from the beginning; (b) with a perturbation signal u(k) and open loop until k = 12.

After the first adaptation it is recommended that the control performance be checked by reference variable changes and forced disturbances. If necessary, some design parameters may be changed.

VIII. SUPERVISION AND COORDINATION

Simulations and practical experience have shown that parameter-adaptive control algorithms work well if all preconditions for stability and convergence are satisfied and if the design factors are chosen properly. Because in practice the preconditions may be violated and the design factors should be

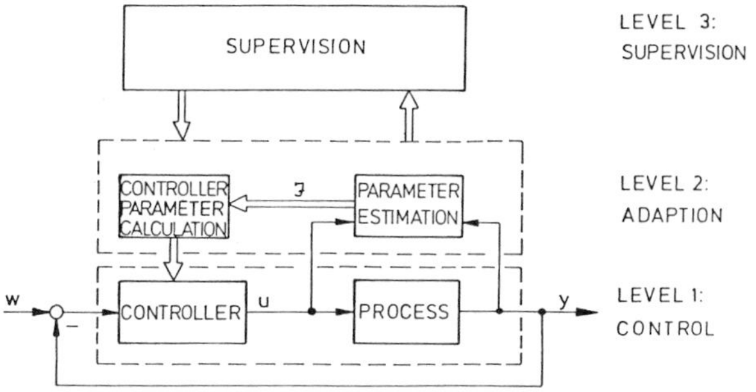

Fig. 3. Block diagram of a parameter-adaptive controller with a supervision level.

altered due to changes of operating conditions, a third feedback level is required to supervise and coordinate the adaptive controller (Schumann *et al.* [21]). See Fig. 3.

A. SUPERVISION

The main elements of PAC, the parameter estimation, and the controller design and the resulting closed loop have to be supervised; that is, possible failures or malfunctions must be detected and diagnosed and appropriate actions must be undertaken.

1. Parameter Estimation

The following possible violations of preconditions are the most common: (1) no persistent excitation, (2) unstationary disturbances, such as jumps and peaks or single-frequency periodic disturbances, (3) fast changes of process statics and dynamics, and (4) wrong model structure parameters (m, d) or sampling time T_0.

As a result of these violations, bad control performance or even unstable control can be observed. A systematic study has shown (Lachmann [27]) that the following quantities particularly are suited for on-line, real-time supervision: (1) a priori error e(k): mean $\overline{e(k)}$ and variance $\hat{\sigma}_e^2(k)$, (2) parameter estimates $\hat{\theta}_i$: variance $\hat{\sigma}^2_{\theta i}$, (3) information matrix $\underline{H}(k) = \underline{P}^{-1}(k)$: trace, and (4) control variable: mean $\overline{y(k)}$ and variance $\hat{\sigma}^2_y(k)$.

An improvement of both the control performance and the supervisory functions could be obtained by the low-pass filtering of the process signals u(k) and y(k) and the process parameters $\hat{\underline{\theta}}(k)$. Dependent on the detected events actions are performed to improve the situation, for example, dynamic parameter estimation stopped for M steps, d.c. value estimation stopped for M steps, restart of parameter estimation (better than change of λ), and automatic search for model order and dead time.

2. Controller Design

Problems may arise from (1) cancellation of unstable poles or zeros, (2) wrong sampling time T_0, and (3) wrong design factors (e.g., r). See Section VII. For the first case, process poles and zeros are calculated and the new controller is only used if there is no (approximate) cancellation.

3. Closed Loop

Possible malfunctions are (1) control error $e_w(k)$ increasing monotonically, (2) actuator position staying at one restriction, and (3) unstable oscillating behavior. In these cases the control system is replaced by a previously designed robust fixed controller. Generally, the closed-loop behavior may be simulated and the new controller is only taken if the control performance becomes better with the new one.

More details on the supervision are given in Isermann and Lachmann [28]. Figure 4 shows two simulation runs with and without supervision. A much better and smoother control performance is achieved by using the various supervisory functions.

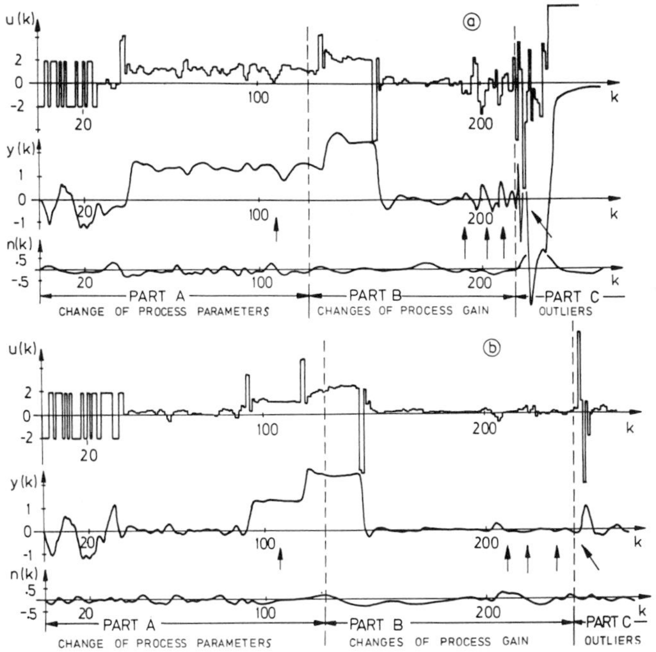

Fig. 4. Parameter-adaptive control (RLS-DB; m = 2; λ = 0.97). Without supervision (top) and with supervision (bottom) for different disturbances.

B. COORDINATION

There are some other functions that should be performed by an adaptive controller. These have to do with the general operating conditions and structural changes within the adaptation level. Some examples are given next.

1. Startup Procedure

If the PAC is just switched on to an unknown process, an undefined control action will result, leading either to a convergent control or to unstable behavior. Therefore, usually, narrow constraints of the manipulated variable are set for the first samples. A better approach is to start open loop with a well-exciting test signal and to switch on the PAC if the identification shows a good verification (comparison of measured and simulated output) or the simulated closed loop shows sufficient performance (Schumann *et al.* [21], Radke [14]).

2. On-Line Search for Model Order and Dead Time

In general it is not difficult to find an appropriate order $\hat{m}$ for the process, because the PAC is not very sensitive within a range $m_0 - 1 \leq \hat{m} \leq m_0 + 2$, where m_0 is the true order. However, the dead time should be known exactly. Therefore an automatic search of $\hat{m}$ and $\hat{d}$ may be implemented. One method based on a recursive calculation of the information matrix $\underline{P}^{-1}(k, \hat{m}, \hat{d})$ is described in Schumann *et al.* [21] and another one based on the nonrecursive representation of the information matrix is given in Mäncher and Hensel [29].

3. Automatic Search for Sampling Time

The choice of an appropriate sampling time T_0 normally is not critical within a certain range, say, for example, 10 to 20 samples per 95% settling time of the transient response. If the sampling time is too small, numerical problems arise for the parameter estimation and the amplitudes of the manipulated variable become too large, especially for deadbeat and minimum variance controllers. For too large a sampling time the control performance becomes poor. Also the absolute value of the settling time and the type of disturbances play a role. Hence, the sampling time T_0 must be selected properly with respect to several considerations.

An automatic search of the sampling time for the controller for a given (small) sampling time of the parameter estimation was described in Schumann *et al.* [21]. Another way is to search for a suitable sampling time based on the simulated control performance during the starting phase.

4. Backup Controller

In order to ensure "fail-safe" operation under any unforeseeable circumstances a robust backup controller (e.g., PI controller) may be designed after the startup procedure.

IX. IMPLEMENTATION ON MICROCOMPUTERS

With respect to the application of PAC, different microcomputer controllers were developed in our laboratory. Two of them (DMR-2, DMR-4) are based on the 8-bit Intel 8080 and 8085 microprocessors with assembler language and one (DMR-16) consists of a 16-bit 8086 microprocessor with PL/M-86 language. They are described in Bergmann [30,31] and Radke [33,14]. Some data are given in Table II.

The main functions of DMR-2 are single-board computer: SBC 80/10, arithmetic processor: AM 9511, console processor: 8748, and PAC: RLS-DB, RLS-MV with explicit d.c. value estimation. The DMR-16 contains single-board computer: SBC 86/12, arithmetic processor: 8087, console processor: 8085, PAC: RLS or DSFI-DB or SC or PID (single-input/single-output), and PAC: RLS-DB or

TABLE II. Technical Data of an 8-Bit and 16-Bit Microcomputer Controller for Adaptive Control

	16 Bit (8086)		*8 Bit (8085)*
Algorithms	*RLS-DB*	*RLS-SC*	*RLS-DB*
Programming language	*PL/M-86*	*PL/M-86*	*Assembler*
Storage			
Code (kbyte)	*4.8*	*10.5*	*3.3*
Data (kbyte)	*1.9*	*7.5*	*1.2*
Execution time (hardware arithmetic) (msec)	*30*	*130*	*90*

DSFI-SC (multiinput/multioutput). Both microcomputers were used several times to test parameter-adaptive control for a variety of industrial processes.

X. APPLICATIONS

The development of the described PAC was accompanied by several applications on our own pilot processes and on industrial processes. A survey of these applications is given in Isermann [32]. The processes are air-conditioning systems, pH-value process, turbulence dryer, dry edge process, and seeds conditioner. Figure 5 shows some examples for an air-conditioning plant.

In Fig. 5a parameter-adaptive PID control is shown for constant airflow. If the PID controller is fixed after the self-tuning phase, the loop comes close to the stability boarder after some setpoint changes (Fig. 5b). Figure 5c demonstrates

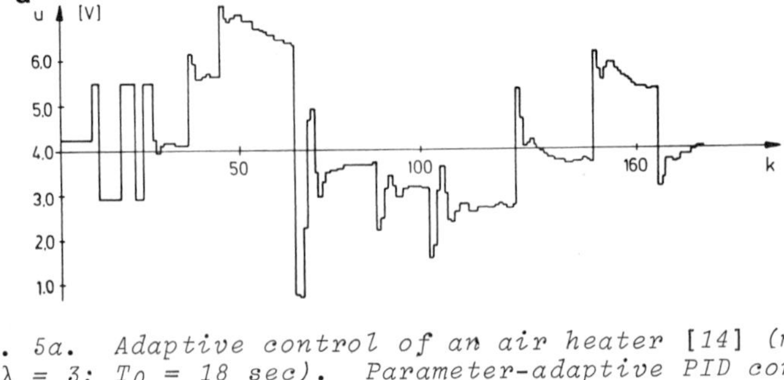

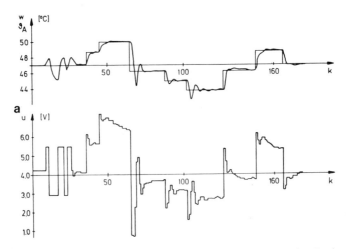

Fig. 5a. Adaptive control of an air heater [14] ($m = 3$; $d = 3$; $\lambda = 3$; $T_0 = 18$ sec). Parameter-adaptive PID control for constant airflow. $\dot{M}_L = 300\ m^3/h$, $r = 0.08$.

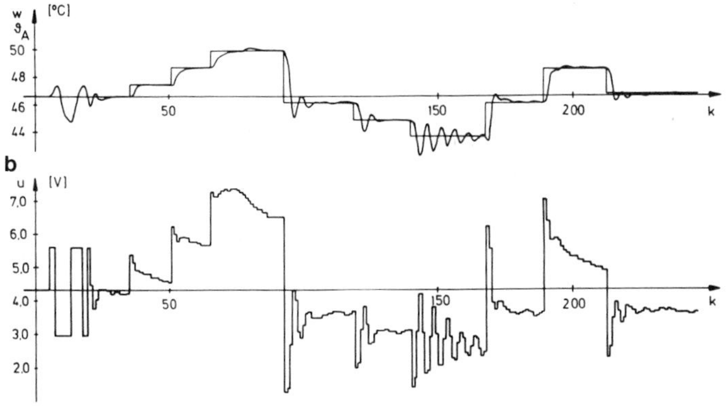

Fig. 5b. Adaptive control of an air heater [14] ($m = 3$; $d = 3$; $\lambda = 3$; $T_0 = 18$ sec). Self-tuning PID control with fixed controller parameters for constant airflow.

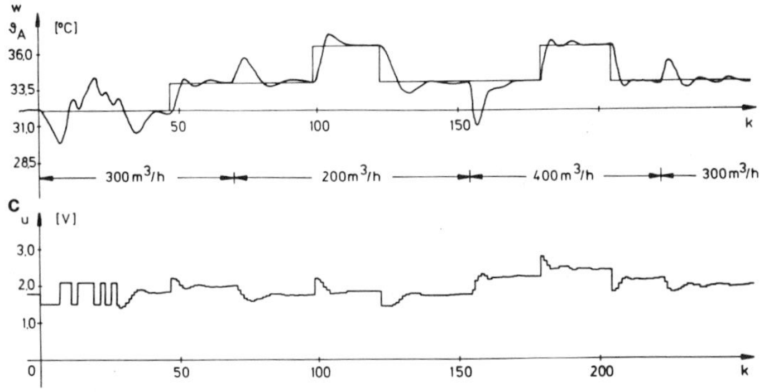

Fig. 5c. Adaptive control of an air heater [14] ($m = 3$; $d = 3$; $\lambda = 3$; $T_0 = 18$ sec). Parameter-adaptive PID control for varying airflow.

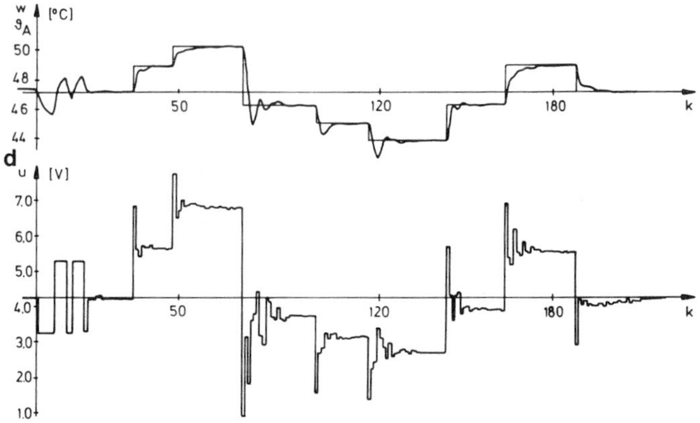

Fig. 5d. Adaptive control of an air heater [14] (m = 3; d = 3; λ = 3; T_0 = 18 sec). Parameter-adaptive state controller for constant airflow.

that the parameter-adaptive PID controller also shows good behavior for airflow disturbances. In Fig. 5d it can be seen that a parameter-adaptive state controller shows approximately the same behavior as the PID controller in Fig. 5a.

The concept of explicit parameter-adaptive control has been extended to *multivariable processes* (Schumann [17,18]) and to several classes of *nonlinear processes* (Lachmann [27]).

XI. CONCLUSIONS

The approaches described to parameter-adaptive controls show that they basically consist of parameter estimation and control algorithms. However, several supplementary functions have to be added to obtain PAC that can be used in practice. This article attempted to summarize the properties of different approaches and to describe improvements in parameter estimation and controller design. There are now mainly adaptive PID and state controllers that show good overall properties. Explicit combinations give a lot of freedom for design, including synchronous and asynchronous controller adaptation, and allow a straightforward extension to multivariable and nonlinear adaptive controllers. The supervision and coordination of the adaptive control level are important for any application. Finally, the implementation on microcomputers and many applications have shown that the parameter-adaptive control is now in a matured status to be used by industry for self-tuning of digital control during the commissioning phase and adaptive control of processes with slowly time variant and nonlinear behavior.

REFERENCES

1. K. J. ÅSTRÖM, "Theory and Applications of Adaptive Control— A Survey," *Automatica 19*, No. 5, 471-486 (1983).
2. R. ISERMANN, "Parameter-Adaptive Control Algorithms— A Tutorial," *Automatica 18*, No. 5, 513-528 (1982).
3. R. E. KALMAN, "Design of a Self-Optimizing Control System," *Trans. ASME 80*, 468-478 (1958).
4. V. PETERKA, "Adaptive Digital Regulation of Noisy Systems," *IFAC Symp. Identif. Process Parameter Estim., 2nd, Prague* (1970).
5. K. J. ÅSTRÖM and B. WITTENMARK, "On Self-Tuning Regulators," *Automatica 9*, 185-199 (1973).
6. D. W. CLARKE and P. J. Gawthrop, "A Self-Tuning Controller," *Proc. IEEE*, 122, 929 (1975).
7. H. KURZ, R. ISERMANN, and R. SCHUMANN, "Development, Comparison and Application of Various Parameter-Adaptive Digital Control Systems," *IFAC Congr., 6th*, Helsinki, June 1978. Revised version in *Automatica 16*, 117-133 (1980).
8. P. E. WELLSTEAD, J. M. EDMUNDS, D. PRAGER, and P. ZANKER, "Self-Tuning Pole/Zero Assignment Regulators," *Int. J. Control*, 1 - 26 (1979).
9. P. EYKHOFF, "System Identification," Wiley, New York, 1974.

10. T. SÖDERSTRÖM, L. LJUNG, and I. GUSTAVSSON, "A Comparative Study of Recursive Identification Methods," Department of Automatic Control, Lund Institute of Technology, Report 7427 (1976).

11. R. ISERMANN, "Digital Control Systems," Springer-Verlag, Berlin and New York, 1977, 1981 (2nd Ed. in German, 1987).

12. V. STREJC, "Least Squares Parameter Estimation," *Automatica 16*, 535 (1980).

13. G. J. BIERMANN, "Factorization Methods for Discrete Sequential Estimation," Academic Press, New York, 1977.

14. F. RADKE, "A Microcomputer System for Testing Parameter-Adaptive Control Systems," Ph.D. Dissertation, TH Darmstadt (in German), Fortschritt-Ber. VDI-Zeitschr., Reihe 8, No. 77, 1984.

15. P. G. KAMINSKI, A. E. BRYSON, and S. F. SCHMIDT, "Discrete Square Root Filtering: A Survey of Current Techniques," *IEEE Trans. Autom. Control AC-16*, 727-736 (1971).

16. I. GUSTAVSSON, L. LJUNG, and T. SÖDERSTRÖM, "Identification of Processes in Closed Loop—Identifiability and Accuracy Aspects, *Automatica 13*, 59 (1977).

17. R. SCHUMANN, "Identification and Adaptive Control of Multivariable Stochastic Linear Systems," *Proc. IFAC Symp. Identif. Syst. Parameter Estim., 5th*, pp. 1203-1212, Pergamon, Oxford, 1979.

18. R. SCHUMANN, "Digital Parameter-Adaptive Multivariable Control," Ph.D. Thesis, TH Darmstadt, Kfk-PDV 217, Kernforschungszentrum, Karlsruhe, 1982.

19. F. RADKE and R. ISERMANN, "A Parameter-Adaptive PID-Controller with Stepwise Parameter Optimization," *Proc. IFAC Congr., Budapest*, Pergamon, Oxford, July 1984, and *Automatica 23*, 1987.

20. R. KOFAHL and R. ISERMANN, "A Simple Method for Automatic Tuning of PID-Controllers Based on Process Parameter Estimation," *Am. Control Conf.*, Boston, Massachusetts, 1985.

21. R. SCHUMANN, K.-H. LACHMANN, and R. ISERMANN, "Towards Applicability of Parameter-Adaptive Control Algorithms," *Proc. IFAC World Congr., 8th, Kyoto, Japan*, Pergamon, Oxford, 1981.

22. R. ISERMANN, "The Application of Parameter-Adaptive Controllers," *Encyl. Syst. Control*, Pergamon, Oxford, 1986.

23. L. LJUNG, "On Positive Real Transfer Functions and the Convergence of Some Recursive Schemes," *IEEE Trans. Autom. Control AC-22*, 539-551 (1977).

24. D. MATKO and R. SCHUMANN, "Self-Tuning Deadbeat Controllers," *Int. J. Control 40*, 393-402 (1984).

25. R. ISERMANN, "Practical Aspects of Process Identification," *Automatica 16*, 575-587 (1980).

26. H. KURZ and W. GOEDECKE, "Digital Parameter-Adaptive Control of Processes with Unknown Constant or Time Varying Deadtime," *Automatica 17*, 245-252 (1981).

27. K.-H. LACHMANN, "Parameter-Adaptive Control Algorithms for Definite Classes of Nonlinear Processes with Unique Nonlinearities," Ph.D. Thesis, TH Darmstadt (in German), Fortschritt-Ber. VDI-Zeitschr., Reihe 8, No. 66, 1983.

28. R. ISERMANN and K.-H. LACHMANN, "Parameter-Adaptive Control with Configuration Aids and Supervision Functions," *Automatica 21*, 625-638, 1985.

29. H. MÄNCHER and H. HENSEL, "Determination of Order and Deadtime for Multivariable Discrete-Time Parameter Estimation Methods," *IFAC Symp. Identif. Syst. Parameter Estim., 7th, York, England*, 1985.

30. S. BERGMANN, "Some Improvements for the Application of Digital Parameter-Adaptive Control Algorithms," *IEEE Conf. Appl. Adapt. Multi-Variable Control, Hull, England*, 1982.

31. S. BERGMANN, "Digital Parameter-Adaptive Control with Microcomputers," Ph.D. Thesis, TH Darmstadt (in German), Fortschritt-Ber. VDI-Zeitschr., Reihe 8, No. 55, 1983.

32. R. ISERMANN, "Computer Aided Design of Digital Control Systems Based on Identified Process Models," *Regelungstechnik*, No. 6, 179-189 and No. 7, 227-234 (1984).

33. F. RADKE, "Implementation of Multivariable Parameter-Adaptive Control Algorithms Using Modern Microprocessor Techniques," *IFAC Symp. Identif. Parameter Estim., 6th, Washington*, June 1982.

Estimation of Transfer Function Models Using Mixed Recursive and Nonrecursive Methods

MOSTAFA HASHEM SHERIF

AT&T Bell Laboratories
Holmdel, New Jersey 07733

LON-MU LIU

Department of Quantitative Methods
University of Illinois at Chicago
Chicago, Illinois 60680

I. INTRODUCTION

Time series analysis, which plays an important role in the understanding and forecasting of stochastic processes, can be performed either in the frequency domain or in the time domain. The frequency domain approach uses properties of the power spectral density, that is, the distribution of the variance of the time series over frequency [1,2]. The time domain approach, in contrast, revolves around correlations of the data at successive time lags.

Nonrecursive algorithms for parameter estimation in the time domain [3,4] are available in many computer programs [e.g., 5-7]. These techniques, however, can become computationally expensive and are not suitable for on-line processing of large quantities of data. Moreover, they are not readily applicable to time-dependent models such as those encountered in business and econometric applications. In contrast, recursive algorithms can track time-varying parameters and reduce the amount of computation needed.

The basis of recursive parameter estimation in time series analysis is to recast the problem into a state estimation framework and then apply nonlinear filtering theory. The components of the state vector may consist of either nonlinear functions of the parameters to be estimated [8-11] or of the parameters themselves [12-17], but it is usually better to form the state vector directly with the parameters. This is because when the state vector is defined as a nonlinear function of the parameters, reparameterization of the model equation could conceal relations among the parameters and obscure the significance of the results. Furthermore, errors could accumulate while the parameters from the state vector were being computed, thereby leading to divergence of the nonlinear filter.

We have previously developed a recursive algorithm for the estimation of Box-Jenkins transfer function models [18]. Results have shown that the algorithm yields accurate estimates even when the system and noise denominator polynomials are different; however, it is sensitive to the choice of initial conditions. This article contains an improved version of our previous algorithm. The basic idea is to obtain the initial values for the recursive algorithm from an early segment of the data with nonrecursive techniques. Using both simulated and real data, we show that the new method compares favorably with traditional maximum likelihood nonrecursive techniques. It is also noted that convergence of the estimates of moving average (MA) parameters in transfer function models could be slower than the convergence of autoregressive (AR) estimates. For recursive estimation, therefore, an autoregressive noise structure should be used for time series with fewer than 200 points.

II. OVERVIEW OF PAST WORK

The focus here will be on three topics, namely, (1) the technique for estimating the system and noise parameters, (2) the method for variance estimation, and (3) the start-up of the iterative procedure. Each item will be discussed separately.

A. ESTIMATION ALGORITHM

Reformulation of the parameter estimation problem within a recursive framework is related to extended Kalman filter (EKF) techniques. Although detailed descriptions of the approach are available in the literature [e.g., 19-20], the Appendix recapitulates the essential points for the reader's convenience.

An autoregressive moving average (ARMA) process includes autoregressive and moving average terms as described by the following equation:

$$y(t) = \phi_1 y(t-1) + \cdots + \phi_p y(t-p) + a_t - \theta_1 a(t-1) - \cdots - \theta_q a(t-q), \tag{1}$$

with $a_t \sim N(0, \sigma_a^2)$. To apply the EKF techniques, (1) should be put in a state-space representation. For example, with additional ϕ_i's and θ_j's, (1) takes the form [8]

$$y(t) = \sum_{i=1}^{r} \phi_i y(t-i) - \sum_{j=0}^{r-1} \theta_j a(t-j), \tag{2}$$

with $\theta_0 = -1$, $r = \max(p, q+1)$, $\phi_i = 0$, $i > p$, and $\theta_j = 0$, $j > q$. To obtain an observable state-space realization, we introduce a state vector $\alpha(t)$ of dimension $r \times 1$ such that

$$\alpha(t) = \left|\begin{array}{c:c} \phi_1 & \\ \phi_2 & I_{r-1} \\ \vdots & \\ \hdashline \phi_r & 0'_{r-1} \end{array}\right| \alpha(t-1) + \left|\begin{array}{c} 1 \\ -\theta_1 \\ \vdots \\ -\theta_{r-1} \end{array}\right| a(t), \quad (3)$$

and the observation equation in the EKF becomes

$$y(t) = z'(t)\alpha(t),$$

where $z(t) = [1, 0'_{r-1}]$, I_{r-1} is the $(r-1) \times (r-1)$ unit matrix, and $0'_{r-1}$ is a row of $r - 1$ zeros. Notice that both the system transition matrix and the noise gain matrix depend on the parameters to be estimated. Likewise, the approaches presented in [9,10] require that the parameters be recomputed from the estimated state vector at each step, and this could lead to the divergence of the parameter estimates.

To alleviate such difficulties, the observation matrix may include the computed residuals $\hat{a}(t)$ so that in the ARMA case we have [12-16]

$$y(t) = [y(t-1) \cdots y(t-p) \quad \hat{a}(t-1) \cdots \hat{a}(t-q)] \times \begin{bmatrix} \phi_1 \\ \vdots \\ \phi_p \\ -\theta_1 \\ \vdots \\ -\theta_q \end{bmatrix} + a(t). \quad (4)$$

With a parameter vector γ of dimension $(p + q)$ defined in terms of the ϕ's and θ's as

$$\gamma = [\phi_1, \ldots, \phi_p, \theta_1, \ldots, \theta_q]', \quad (5)$$

(4) takes the form

$$y(t) = H(t)\gamma(t) + a(t),$$

where

$$H(t) = [y(t-1) \cdots y(t-p) \quad \hat{a}(t-1) \cdots \hat{a}(t-q)]. \tag{6}$$

Multidimensional ARMAs and nonstationary models can be similarly treated [11,16]. Consider, for example, the autoregressive-integrated moving average (ARIMA) model given by

$$\phi(B)(1 - B)^d y(t) = \theta(B)a(t), \tag{7}$$

where B is the lag operator such that $By(t) \triangleq y(t-1)$; $\phi(B) = 1 - \phi_1 B - \cdots - \phi_p B^p$; $\theta(B) = 1 - \theta_1 B - \cdots - \theta_q B^q$; and d the number of differences needed to obtain a stationary signal. To apply the procedure defined by (4)-(6), it is sufficient to rewrite (7) as

$$\Phi(B)y(t) = \theta(B)a(t),$$

with

$$\Phi(B) = \phi(B)(1 - B)^d.$$

Transfer function models consist of two parts, the system component and the noise component. Although it is possible to estimate both the system and noise parameters recursively in a sequential way [14,22], real-time applications require their simultaneous estimation. Panuska [21] provided an algorithm for multi-input transfer function models under the assumption that the noise autoregressive structure is identical to the denominator polynomial of the system. That is, the system has the structure

$$y(t) = \frac{\omega(b)}{\delta(b)} x(t) + \frac{\theta(B)}{\phi(B)} a(t) \quad \text{with} \quad \delta(B) = \phi(B),$$

which is similar to the structure assumed for the recursive maximum likelihood (RML) method [14]. Panuska's method, however, suffers from several limitations. It is not always realistic to assume that the system and noise denominator polynomials are identical, and when they differ, the equations in his method become highly nonlinear and involve products of parameters. It is not clear how the algorithm would behave in these instances since results with this approach do not seem to be readily available.

A general recursive algorithm should satisfy the following requirements: (1) it should be applicable even when the system and noise parameters are different, (2) it should estimate the system and noise parameters simultaneously, and (3) it should minimize the need for guessing the initial values of the recursion. It will be seen that the algorithm described in Section III, which is an improved version of the algorithm in [18], fulfills the three conditions.

B. ESTIMATION OF NOISE VARIANCE

There are several methods for estimating the noise variance, namely:

1. The parameter vector γ and the noise variance σ_a^2 may be estimated recursively through an adaptive limited memory filter that minimizes the quadratic loss function [23]:

$$L(\gamma) = \sum_{k=0}^{t-1} [y(t - k) - H(t - k)\hat{\gamma}(t - k)]^2 \lambda^k, \tag{8}$$

where λ, $0 < \lambda < 1$, is a weighting factor that emphasizes the most recent observation.

2. The residuals variance may be estimated from the predicted observation error $r(t)$ defined as

$$r(t) = y(t) - H(t)\hat{\gamma}(t|t - 1)$$

by the formula [24]

$$\hat{\sigma}_a^2(t) = \frac{1}{t-1}\left\{\sum_{j=1}^{t} (r(j) - \bar{r})^2 - \frac{t-1}{t} \times [H(t)P(t|t-1)H'(t) - P(t|t)]\right\}, \tag{9}$$

where $\hat{\gamma}(t|t-1)$ is the prediction of $\gamma(t)$ based on all observations up to $(t-1)$, $P(t|t-1)$ is the predicted covariance of the parameter estimates as obtained from the Kalman filter, and $\bar{r} = (1/t)\Sigma_{j=1}^{t} r(t)$. Note that the estimate from (9) could become negative due to accumulated numerical errors.

3. If the filter has run long enough and the steady state has been reached, other techniques could be used [25-27]. These algorithms, however, require complex computations and are not suitable for time series with fewer than, say, 1000 points. The Alspach algorithm [27], in particular, is based on a Bayesian procedure and is computationally demanding.

4. The recursive maximum likelihood estimator given by [21]

$$\hat{\sigma}_a^2(t) = \hat{\sigma}_a^2(t-1) + (1/t)[r^2(t) - r^2(t-1)], \tag{10}$$

where $r(t)$ is the predicted error, will yield biased estimates.

5. A fictitious noise with covariance Ω may be introduced in (5) to represent possible fluctuations in the parameters and/or modeling errors. The appropriate noise levels can be determined by trial and error through off-line simulations to ensure a satisfactory performance [19, pp. 305-307]. Alternative approaches based on maximizing a likelihood function are available [8,17,28, pp. 364-368]. For real-time operation, Ω may be estimated as in (9) [24,29], but potential numerical instabilities with this method could lead to negative variance estimates.

C. *START-UP PROCEDURE*

The computation of the missing starting values for the noise residuals $\hat{a}(t)$ poses significant difficulties in nonrecursive algorithms [3,30-32]. These difficulties increase in recursive estimation because the initial values of many parameters are also unknown. Furthermore, the early residuals could deflect the estimate trajectory, and for short time series, they could prevent convergence within the time frame of interest [13,18].

There are several avenues to overcome the previous problems. It is possible to delay parameters update by about 20 to 30 points [16,18,24] and/or to postpone the start-up of the iterative procedure itself [16, p. 50]. Szelag [33] proposed a third approach for short-term forecasting of telephone trunks demand. He divided a time series of 33 points into two subseries of 16 and 17 points, respectively, and obtained the initial conditions for the recursive filter from the first segment through a least-squares recursive algorithm. The algorithm presented in the following section will incorporate all these ideas.

III. THE NEW ALGORITHM

To facilitate the presentation, we shall follow the same outline used in the preceding section.

A. *PARAMETER ESTIMATION*

Consider the following general multiple-input transfer function model

$$y(t) = C + \frac{\omega_1}{\delta_1(B)} x_1(t) + \cdots + \frac{\omega_m}{\delta_m(B)} x_m(t) + \frac{\theta(B)}{\phi(B)} a(t), \quad t = 1, 2, \ldots, n, \tag{11}$$

where $\{y(t)\}$ is the stationary series of the deviation of the output from its mean; $\{x_l(t)\}$ is the system stationary input t that is independent of the noise $a(t)$, for all values of l where $l = 1, \ldots, m$; $\{a(t)\}$ is the Gaussian noise distributed as $N\left(0, \sigma_a^2\right)$; C is a constant term related to the means of the output series and of the input series $\{x_l(t), l = 1, \ldots, m\}$;

$$\phi(B) = 1 - \phi_1 B - \phi_2 B^2 - \cdots - \phi_p B^p, \tag{12a}$$

$$\theta(B) = 1 - \theta_1 B - \theta_2 B^2 - \cdots - \theta_q B^q; \tag{12b}$$

and $\omega(B)$ and $\delta(B)$ have the general form

$$\omega(B) = \omega_0 + \omega_1 B + \cdots + \omega_s B^s, \tag{12c}$$

$$\delta(B) = 1 - \delta_1 B - \cdots - \delta_r B^r. \tag{12d}$$

We assume a unidirectional relationship between the inputs and output, that is, $\{x_l(t), l = 1, \ldots, m\}$ are independent of $\{y(t)\}$ and there is no feedback. Thus, $x_l(t)$ and $a(t)$ are independent for all $l = 1, \ldots, m$.

Consider first the single-input model

$$y(t) = \frac{\omega(B)}{\delta(B)} x(t) + \frac{\theta(B)}{\phi(B)} a(t), \qquad t = 1, 2, \ldots, n. \tag{13}$$

The parameter vector $\gamma(s + r + p + q + 1)$, defined as

$$\gamma[\phi_1 \cdots \phi_p, \theta_1 \cdots \theta_q, \omega_0 \cdots \omega_s, \delta_1 \cdots \delta_r]',$$

is assumed to evolve according to the equation

$$\gamma(t) = T\gamma(t - 1) + \epsilon(t), \qquad \epsilon(t) \sim N(0, \Omega), \tag{14}$$

where T is an appropriate transition matrix. Another assumption is that $\epsilon(t)$ and $a(t)$ are independent. One fundamental difference between this formulation and conventional Box-Jenkins analysis is that the model parameters may change over time according to a vector AR(1) model.

In most of the engineering literature, the transition matrix T has a physical meaning; if unknown, however, it can be estimated by maximum likelihood methods [17]. We will consider the case where T is the identity matrix and $\Omega = 0$ (i.e., the model is linear and time invariant) to compare the performance of the algorithm with the nonrecursive method of Box and Jenkins [3].

We now rewrite (13) as

$$y(t) = \left[1 - \frac{\phi(B)}{\theta(B)}\right] y(t) + \frac{\omega(B)}{\delta(B)} \frac{\phi(B)}{\theta(B)} x(t) + a(t)$$

$$= h(\gamma(t - 1), Y_{t-1}, X_t) + a(t), \qquad (15)$$

where Y_{t-1} denotes the set $\{y(t - 1), y(t - 2), \ldots\}$ and X_t the set $\{x(t), x(t - 1), \ldots\}$. The likelihood function of the data given by the Bayes rule is

$$P(Y_t | X_t, \gamma(i)) = \prod_{i=1}^{t} P(Y(i) | Y_{i-1}, X_i; \gamma(i - 1))$$

$$= \prod_{i=1}^{t} P(\hat{a}(i)),$$

with

$$\hat{a}(t) = y(t) - h(Y_{t-1}, X_i; \hat{\gamma}(t - 1)). \qquad (16)$$

It is possible to derive maximum likelihood estimates of γ in the Gaussian case [34, pp. 91-95], but the results depend on whether the common variance of the innovations is known. One advantage of the EKF algorithm is that it avoids this difficulty.

To derive the EKF equations, we expand (15) around a reference trajectory $\hat{\gamma}(t - 1)$ as follows:

$$y(t) \cong h(\hat{\gamma}(t-1), Y_{t-1}, X_t) + \left.\frac{\partial h(\cdot)}{\partial \gamma}\right|_{\gamma=\hat{\gamma}(t-1} \Delta\gamma(t) + a(t)$$

$$= h(\hat{\gamma}(t-1), Y_{t-1}, X_t) + \sum_{j=1}^{p} v_j(t)(\phi_j(t) - \hat{\phi}_j(t-1))$$

$$+ \sum_{j=1}^{q} w_j(t)\Big(\theta_j(t) - \hat{\theta}_j(t-1)\Big)$$

$$+ \sum_{j=0}^{s} g_j(t)\Big(\omega_j(t) - \hat{\omega}_j(t-1)\Big)$$

$$+ \sum_{j=1}^{r} k_j(t)\Big(\delta_j(t) - \hat{\delta}_j(t-1)\Big) + a(t) \tag{17}$$

or

$$\Delta y(t) = H(t|t-1)\ \Delta\gamma(t) + a(t),$$

where

$$\begin{aligned} H(t|t-1) = [&v_1(t|t-1) \cdots v_p(t|t-1) w_1(t|t-1) \\ &\cdots w_q(t|t-1) g_0(t|t-1) \\ &\cdots g_s(t|t-1) k_1(t|t-1) \cdots k_r(t|t-1)], \end{aligned} \tag{18}$$

$$\begin{aligned} v_j(t) &= \left.\frac{\partial h(\gamma(t-1), Y_{t-1}, X_t)}{\partial \phi_j}\right|_{\gamma=\hat{\gamma}(t-1)}, \\ &\qquad j = 1, \ldots, p, \\ w_j(t) &= \left.\frac{\partial h(\gamma(t-1), Y_{t-1}, X_t)}{\partial \theta_j}\right|_{\gamma=\hat{\gamma}(t-1)}, \\ &\qquad j = 1, \ldots, q, \\ g_j(t) &= \left.\frac{\partial h(\gamma(t-1), Y_{t-1}, X_t)}{\partial \omega_j}\right|_{\gamma=\hat{\gamma}(t-1)}, \\ &\qquad j = 0, \ldots, s, \end{aligned} \tag{19}$$

$$k_j(t) = \left.\frac{\partial h(\gamma(t-1), Y_{t-1}, X_t)}{\partial \delta_j}\right|_{\gamma=\hat{\gamma}(t-1)},$$

$$j = 1, \ldots, r.$$

The recursive equations to estimate the v_j's, w_j's, g_j's, and k_j's are

$$\hat{\phi}(B)v_j(t) = \hat{a}(t-j), \tag{20a}$$

$$\hat{\theta}(B)w_j(t) = -\hat{a}(t-j), \tag{20b}$$

$$\hat{\delta}(B)\hat{\theta}(B)g_j(t) = \hat{\phi}(B)x(t-j), \tag{20c}$$

$$\hat{\delta}^2(B)\hat{\theta}(B)k_j(t) = \hat{\omega}(B)\hat{\phi}(B)x(t-j). \tag{20d}$$

When $\delta(B) = \phi(B)$ (i.e., in Panuska's model), these equations reduce to

$$\begin{aligned} \hat{\theta}(B)g_j(t) &= x(t-j), \\ \hat{\delta}(B)\hat{\theta}(B)k_j(t) &= \hat{\omega}(B)x(t-j). \end{aligned} \tag{21}$$

In multiple-input transfer function models, (20c) and (20d) for the ith input variable become

$$\hat{\delta}_i(B)\hat{\theta}(B)g_{ij}(t) = \hat{\phi}(B)x_i(t-j), \tag{22a}$$

$$\hat{\delta}_i^2(B)\hat{\theta}(B)k_{ij}(t) = \hat{\omega}_i(B)\hat{\phi}(B)x_i(t-j). \tag{22b}$$

The constant C, if present in (11), is treated as the gain for an input series of {1}. For a single-input transfer function model with a constant term we thus have

$$y(t) = \frac{\omega_1(B)}{\delta_1(B)} x_1(t) + \frac{w_2(B)}{\delta_2(B)} x_2(t) + \frac{\theta(B)}{\phi(B)} a(t) \tag{23}$$

with $\omega_1(B) = C$, $\delta_1(B) = 1$, and $x_1(t) = [1, 1, \ldots]'$, and the update equations are

$$\hat{\phi}(B)v_j(t) = \hat{a}(t-j),$$

$$\hat{\theta}(B)w_j(t) = -\hat{a}(t-j),$$

$$\hat{\theta}(B)g_{10}(t) = \hat{\phi}(B),$$

$$\hat{\delta}(B)\hat{\theta}(B)g_{2j}(t) = \hat{\phi}(B)x_2(t - j), \tag{24}$$

$$\hat{\delta}^2(B)\hat{\theta}(B)k_{2j}(t) = \hat{\omega}(B)\hat{\phi}(B)x_2(t - j).$$

The EKF equations are given by (see Appendix)

$$\hat{\gamma}(t|t - 1) = T\hat{\gamma}(t - 1|t - 1), \tag{25a}$$

$$P(t|t - 1) = TP(t - 1|t - 1)T' + \Omega, \tag{25b}$$

$$K(t) = P(t|t - 1)H'(t|t - 1) \times \left[H(t|t - 1)P(t|t - 1)H'(t|t - 1) + \sigma_a^2\right]^{-1}, \tag{25c}$$

$$\hat{\gamma}(t|t) = \hat{\gamma}(t|t - 1) + K(t)[\hat{a}(t|t)], \tag{25d}$$

$$P(t|t) = [I - K(t)H(t|t - 1)]P(t|t - 1) \times [I - K(t)H(t|t - 1)]' + \sigma_a^2 K(t)K'(t). \tag{25e}$$

B. ESTIMATION OF VARIANCE

From the definition of the residual $\hat{a}(t)$ in (6) we have

$$\begin{aligned}\hat{a}(t) &= y(t) - H(t|t - 1)\hat{\gamma}(t|t) \\ &= H(t|t - 1)[\gamma(t) - \hat{\gamma}(t|t - 1)] + a(t).\end{aligned} \tag{26}$$

Therefore,

$$\begin{aligned}E[\hat{a}^2(t)] &= H(t|t - 1)P(t|t)H'(t|t - 1) + \sigma_a^2(t) \\ &\quad - H(t|t - 1)K(t)\sigma_a^2(t) - \sigma_a^2(t)K'(t)H'(t|t - 1) \\ &= H(t|t - 1)P(t|t)H'(t|t - 1) + \sigma_a^2(t) \\ &\quad \times \left[H(t|t - 1)P(t|t - 1)H'(t|t - 1) + \sigma_a^2(t)\right]^{-1}\sigma_a^2(t) \\ &\quad - H(t|t - 1)P(t|t - 1)H'(t|t - 1) \\ &\quad \times \left[H(t|t - 1)P(t|t - 1)H'(t|t - 1) + \sigma_a^2(t)\right]^{-1}\sigma_a^2(t).\end{aligned} \tag{27}$$

Define the parameters α, β, and η such that

$$H(t|t-1)P(t|t)H'(t|t-1) = \alpha,$$

$$H(t|t-1)P(t|t-1)H'(t|t-1) = \beta,$$

and

$$E[\hat{a}^2(t)] = \frac{1}{m}\sum_{t=1}^{m} \hat{a}^2(t) = \eta.$$

The quadratic equation in (27) has a positive root:

$$\sigma_a^2(t) = \frac{1}{2}\left[\eta + \sqrt{\eta^2 - 4(\eta - \alpha)\beta}\right]. \tag{28}$$

Near convergence, the values of α and β are close to zero and (28) reduces to the result in [18]:

$$\hat{\sigma}_a^2(t) \cong \frac{1}{m}\sum_{t=1}^{m} \hat{a}^2(t) = \eta. \tag{29}$$

Both (28) and (29) can take a recursive form similar to (10); however, (28) is more useful for short time series since (29) tends to overestimate the variance at the early points.

To de-emphasize earlier residuals and to allow for time-varying variances, $E[\hat{a}^2(t)]$ is estimated using the most recent m (say, 25 or 50) observations. When the number of data points available is smaller than m, the estimated variance $\hat{\sigma}_a^2$ is a weighted average of the initial $\hat{\sigma}_a^2$ and the available $\hat{a}(t)$'s.

If we ignore the problems associated with starting values for the time being, the steps in the algorithm are as follows:

1. Update the matrix H using (18) and (19).
2. Obtain new estimates using the EKF described in (25).
3. Calculate the residuals using (16).
4. Update the variance estimator using (28).
5. Update Ω (if needed).
6. Move to the next iteration.

C. START-UP PROCEDURE

In a typical Kalman filter application, the starting values for $a(t)$, $v_j(t)$, $w_j(t)$, $g_j(t)$, and $k_j(t)$ are set to zero and the initial covariance matrix $P(0)$ is diagonal with large entries [18]. The initial values of the state vector are selected to ensure convergence of the estimates. Unfortunately, the use of zero starting values for $g_j(t)$ and $k_j(t)$ in (20) may perturb the estimation of the moving average parameters [35]. Accordingly, the first few values (say, 10) of the recursive computations of $g_j(t)$ and $k_j(t)$ should be discarded to ensure that start-up effects have died out [35], and the update of $P(t|t)$ will begin at the 10th data point. Notice that the parameters update starts at the 51st point and that the initial values for the recursive estimation are obtained through a nonrecursive maximum likelihood algorithm with the series formed with the first 50 data points.

D. FILTER CONVERGENCE

To ensure the filter convergence, a projection mechanism confines the estimates to the region bounded by the unit circle [3]. Ljung's convergence theory [36,37] requires that the input $\{x(t)\}$ be a weakly stationary process with a rational spectral density, that all absolute moments of the sequence $\{x(t)\}$ and $\{a(t)\}$ exist and are bounded, and that the original system of (13) and the innovation (15) are stable. It can be shown that these conditions are satisfied [18].

IV. EXAMPLES

This section contains results from two sets of time series. The first data set consists of a simulated time series, while the second time series is that of the sales and leading indicator data published by Box and Jenkins as series M [3].

A. SIMULATED TIME SERIES

Consider the system

$$y(t) = C + \frac{\omega B}{1 - \delta B} x(t) + (1 - \theta B)a(t), \quad t = 1, 2, \ldots, n.$$

The series y(t) and x(t) are generated with the parameters $C = 1.5$, $\omega = 2.7$, $\delta = 0.7$, $\theta = 0.5$, $x(t) \sim N(0, 9)$, $a(t) \sim N(0, 1)$, and $n = 500$. From (20), the update equations are

$$w_1(t) = \hat{\theta} w_1(t - 1) - \hat{a}(t - 1),$$

$$g_{10}(t) = \hat{\theta} g_{10}(t - 1) + 1,$$

$$g_{21}(t) = \frac{x(t - 1)}{(1 - \hat{\delta}B)(1 - \hat{\theta}B)},$$

$$k_{21}(t) = \frac{\hat{\omega} x(t - 2)}{(1 - \hat{\delta}B)^2 (1 - \hat{\theta}B)}.$$

To estimate the residuals we use the relation

$$\hat{a}(t) = \frac{y(t)}{1 - \hat{\theta}B} - \frac{\hat{C}}{1 - \hat{\theta}B} - \frac{\hat{\omega} x(t - 1)}{(1 - \hat{\theta}B)(1 - \hat{\delta}B)}$$

and the variance is calculated from (28) with $m = 50$.

Table I contains the parameter estimates for two sets of initial values, and Fig. 1 depicts the convergence plots for the first case. Clearly, the mixture approach yields good estimates for the system parameters (C, ω, and δ), but, as suggested by previous investigations [31,35], the estimate of the moving average parameter (θ) is less stable. This is because in a moving average (MA) process, the residual at each instant depends on all previous values of the residuals. The starting values for a(t) and the initial values of the parameters will

TABLE I. Parameters Estimates for Various Initial Conditions Using Mixture Method[a]

Parameter	Case I Initial value[b]	Case I Final value	Case II Initial value[b]	Case II Final value
C	1.73	1.531	1.73	1.535
ω	2.73	2.757	2.73	2.757
δ	0.70	0.699	0.70	0.699
θ	0.12	0.419	0.35	0.419
σ_a^2	1.80	1.051	1.80	1.057

[a] *Model:*

$$y(t) = C + \frac{\omega B}{1 - \delta B} x(t) + (1 - \theta B) a(t),$$

$$C = 1.5, \quad \omega = 2.75, \quad \delta = 0.7, \quad \theta = 0.5,$$

$$x \sim N(0, 9), \quad a \sim N(0, 1).$$

[b] *Obtained from the first 50 points of the data by nonrecursive maximum likelihood estimation* [7].

have a lasting influence on the entire residual series and will affect the convergence of the MA estimates. In contrast, when the noise component follows an autoregressive (AR) process, effects of each residual are confined to a finite duration. Estimates of the AR parameters are therefore less sensitive to biases caused by starting values and initial conditions. Consequently, better fits could be obtained by replacing a moving average noise with an equivalent autoregressive process.

Although an MA(1) process is equivalent to an infinite AR process, as a first approximation, we replace the MA(1) noise with an AR(1) process. Thus we will have the model

$$y(t) = C + \frac{\omega B}{1 - \delta B} x(t) + \frac{1}{1 - \phi B} a(t).$$

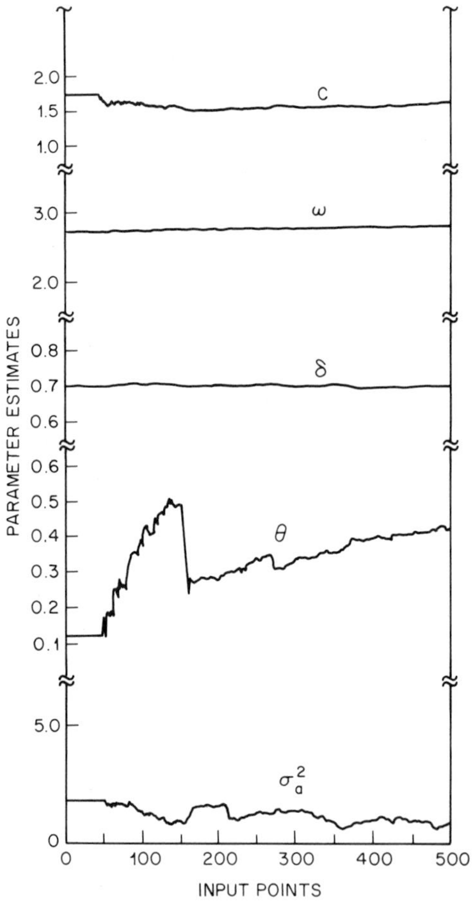

Fig. 1. Convergence plots for the simulated model using the mixture method.

The update equations for the approximate AR(1) model are

$$v_1(t) = \hat{\phi} v_1(t - 1) + \hat{a}(t - 1),$$

$$g_{10}(t) = 1 - \hat{\phi}B,$$

$$g_{21}(t) = \frac{(1 - \hat{\phi}B)x(t - 1)}{1 - \hat{\delta}B},$$

$$k_{21}(t) = \frac{\hat{\omega}(1 - \hat{\phi}B)x(t - 1)}{(1 - \hat{\delta}B)^2}.$$

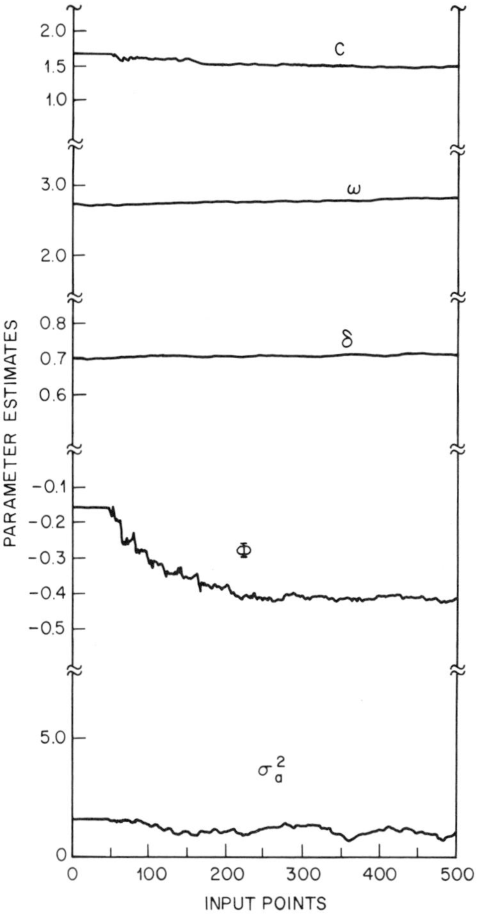

Fig. 2. Convergence plots for the simulated model with autoregressive noise assumed for the fit.

The residuals are computed from the formula

$$\hat{a}(t) = (1 - \hat{\phi}B)(y(t) - \hat{C}) - \frac{(1 - \hat{\phi}B)\hat{\omega}x(t - 1)}{1 - \hat{\delta}B},$$

and the variance is calculated as before. The results depicted in Fig. 2 and summarized in Table II confirm the superiority of the approximate AR model to the corresponding MA process in terms of convergence properties.

TABLE II. Parameters Estimates for Simulated Data (Noise Modeled as an Autoregressive Process)[a]

Parameter	*Initial value*[b]	*Final value*
C	1.68	1.543
ω	2.73	2.749
δ	0.70	0.699
ϕ	-0.16	-0.421
σ_a^2	1.60	1.025

[a] *Actual model:*

$$y(t) = C + \frac{\omega B}{1 - \delta B} x(t) + (1 - \theta B)a(t).$$

Fitted model:

$$y(t) = C + \frac{\omega B}{1 - \delta B} x(t) + \frac{1}{1 - \phi B} a(t),$$

$$C = 1.5,\ \omega = 2.75,\ \delta = 0.7,\ \theta = 0.5,$$

$$x \sim N(0, 9),\ a \sim N(0,1).$$

[b] *Obtained from the first 50 points of the data by nonrecursive maximum likelihood estimation [7].*

B. SALES AND LEADING INDICATOR DATA

To investigate the algorithm performance with real data, we consider the sales and leading indicator series listed as series M in Box and Jenkins [3, p. 537]. The data, shown in Fig. 3, consist of 150 pairs of observations. After differencing the original x and y series, Box and Jenkins proposed the following model to describe the relation between the differenced x and y series [3]:

$$y(t) = C + \frac{\omega_3 B^3 x(t)}{1 - \delta B} + (1 - \theta B)a(t).$$

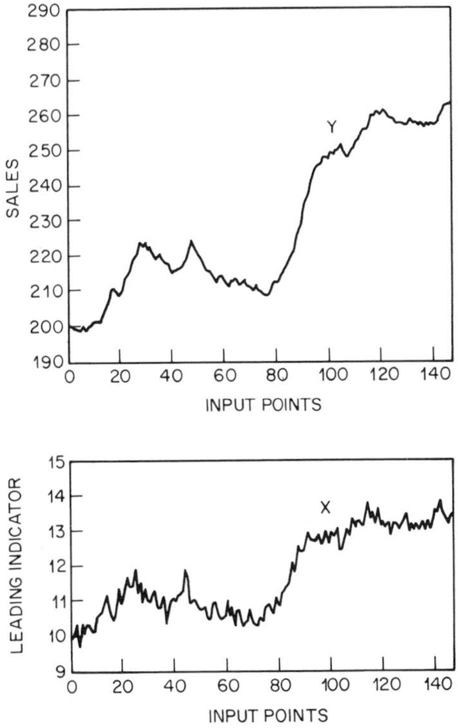

Fig. 3. Sales and leading indicator data from Box and Jenkins [3].

The update equations are

$$w_1(t) = \hat{\theta} w_1(t - 1) - \hat{a}(t - 1),$$

$$g_{10}(t) = \hat{\theta} g_{10}(t - 1) + 1,$$

$$g_{23}(t) = \frac{x(t - 3)}{(1 - \hat{\delta}B)(1 - \hat{\theta}B)},$$

$$k_{21}(k) = \frac{\omega_3 x(t - 4)}{(1 - \hat{\delta}B)^2(1 - \hat{\theta}B)}.$$

The residuals are calculated from

$$a(t) = \frac{y(t)}{1 - \hat{\theta}B} - \frac{C}{1 - \hat{\theta}B} - \frac{\omega_3 x(t - 3)}{(1 - \hat{\delta}B)(1 - \hat{\theta}B)},$$

and the variance is computed from (28) with $m = 50$.

TABLE III. Estimates for Sales and Leading Indicator Data

Parameter	Mixture method: Initial value[a]	Mixture method: Final value	SCA
C	0.040	0.034	0.034(0.078)[b]
ω	4.700	4.650	4.70(0.048)
δ	0.720	0.727	0.735(0.004)
θ	0.280	0.494	0.688(0.066)
σ_a^2	0.050	0.036	0.046

[a] *Obtained from the first 50 points of the data by nonrecursive maximum likelihood estimation* [7].

[b] *Standard error.*

Table III contains the parameter estimates for different initial conditions with the new algorithm and the estimates from the maximum likelihood algorithm in the SCA system [7]. The convergence plots of Fig. 4 confirm that estimates of the parameters C, ω, and δ are satisfactory while the convergence of the MA estimate is slow. Furthermore, the example illustrates how recursive estimation can provide new insights into the data. The plots indicate that the estimate for the "constant" term is stable only within the interval 80-110, that is, when there is a steep rise in the sales data. Nevertheless, the low value of the estimate of C (around 0.02) suggests that this term may not be statistically significant. The concomitant increase in the residuals variance seems to offer additional support to the hypothesis that the proposed model is deficient in this interval.

Since the difficulty of estimating the moving average parameter in short series can be avoided by using an equivalent

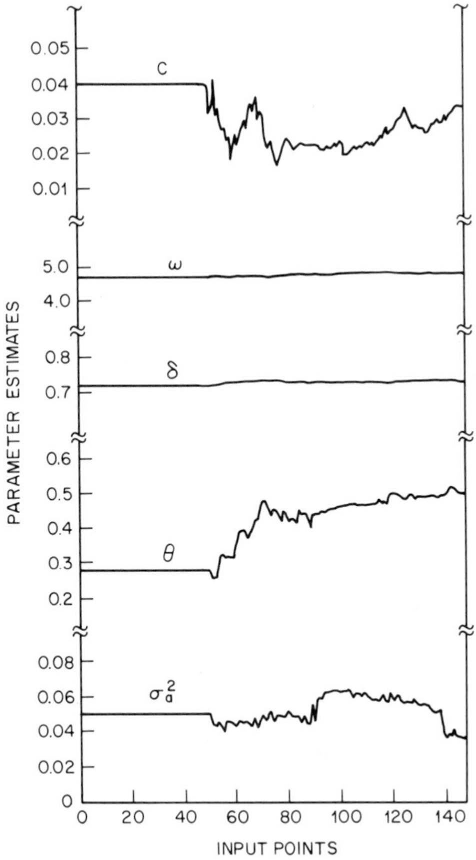

Fig. 4. Convergence plots for the sales and leading indicator data with moving average noise. Initial values for the recursive estimation are obtained from the first 50 points through a nonrecursive maximum likelihood algorithm [7].

autoregressive process, we utilize the following approximate model for this set of data:

$$y(t) = C + \frac{\omega_3 B^3 x(t)}{1 - \delta B} + \frac{1}{1 - \phi B} a(t).$$

Here the update equations are

$$v_1(t) = \hat{\phi} v_1(t - 1) + \hat{a}(t - 1)$$

$$g_{10}(t) = 1 - \hat{\phi}B,$$

$$g_{23}(t) = \frac{(1 - \hat{\phi}B)x(t - 3)}{1 - \hat{\delta}B},$$

$$k_{21}(t) = \frac{\hat{\omega}_3(1 - \hat{\phi}B)x(t - 4)}{1 - \hat{\delta}B}.$$

and the residuals are given by

$$\hat{a}(t) = (1 - \hat{\phi}B)(y(t) - \hat{C}) - \frac{(1 - \hat{\phi}B)\hat{\omega}_3 x(t - 3)}{1 - \hat{\delta}B}.$$

Results of the estimation are given in Table IV along with the maximum likelihood estimates obtained using the SCA system [7]. The convergence plots of Fig. 5 confirm that, with an autoregressive noise, the initial transients subside more rapidly and the speed of convergence of the noise estimates is improved. On the basis of this analysis, it is recommended that an autoregressive noise model be used instead of a moving average model for time series with fewer than 200 points.

TABLE IV. Parameter Estimates for Leading Indicator Data (Noise Modeled as an Autoregressive Process)

	Mixture method		
Parameter	*Initial value*[a]	*Final value*	*SCA*
C	0.040	0.033	0.036(0.02)[b]
ω	4.800	4.717	4.744(0.06)
δ	0.720	0.724	0.723(0.01)
ϕ	-0.190	-0.405	-0.446(0.08)
σ_a^2	0.050	0.038	0.051

[a] *Obtained from the first 50 points of the data by nonrecursive maximum likelihood estimation [7].*

[b] *Standard error.*

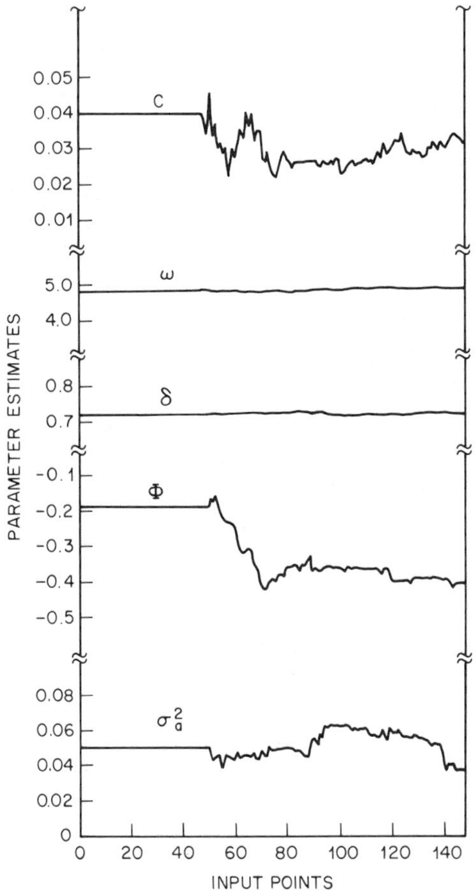

Fig. 5. Convergence plots for sales and leading indicator data with autoregressive noise assumed for the fit. Initial values for the recursive estimation are obtained from the first 50 points through a nonrecursive maximum likelihood algorithm [7].

V. SUMMARY

Selection of suitable initial values for the recursive estimation of Box-Jenkins transfer function models is not always easy because the choice depends on individual experience and requires a good knowledge of the system characteristics. Furthermore, for complex models, improper specification could result in poor performance and slower convergence. In this paper, we investigated the use of combined nonrecursive and recursive estimation techniques so that a nonrecursive algorithm, based on maximum likelihood estimation, could be used to derive the initial state vector for the recursive estimation from a subset of the data.

The results indicate that, for an autoregressive noise, the estimates obtained with the mixture method compare favorably with estimates using traditional nonrecursive algorithms. With a moving average noise, however, convergence could be delayed. Therefore, it is suggested that moving average noise be replaced with an equivalent autoregressive noise during the recursive estimation of the parameters of short time series (fewer than 200 points).

VI. APPENDIX

For the extended Kalman filter (EKF), let a nonlinear discrete time system be given by

$$\gamma(t) = f(t, \gamma(t-1)) + \epsilon(t), \quad t = 1, 2, \ldots, n, \tag{A1}$$

$$y(t) = h(t, \gamma(t)) + a(t), \quad t = 1, 2, \ldots, n, \tag{A2}$$

where $\gamma(t)$ is the state vector, $y(t)$ is the observation vector, $\epsilon(t)$ and $a(t)$ are noises with covariance matrices $\Omega(t)$ and $\Sigma(t)$, respectively.

The EKF estimates the state vector at time t based upon the present and past observations according to the recursive relation

$$\hat{\gamma}(t|t) = f(t, \hat{\gamma}(t|t-1)) + K(t)[y(t) - h(t, \hat{\gamma}(t|t-1))], \tag{A3}$$

where $\hat{\gamma}(t_1|t_2)$ is the estimate of the state vector γ at time t_1 based on the observations up to t_2 and $K(t)$ is the Kalman gain.

The following relations define the Kalman fitler:

$$K(t) = P(t|t-1)H'(t, \hat{\gamma}(t|t-1)) \times [H(t, \hat{\gamma}(t|t-1))P(t|t-1)H'(t, \hat{\gamma}(t|t-1)) + \Sigma(t)]^{-1}, \quad \text{(A4)}$$

$$P(t|t-1) = F(t, \hat{\gamma}(t-1))P(t-1|t-1)F'(t, \hat{\gamma}(t-1)) + \Omega(t), \quad \text{(A5)}$$

$$P(t|t) = [I - K(t)H(t, \hat{\gamma}(t|t-1))] \times P(t|t-1)[I - K(t)H(t, \hat{\gamma}(t|t-1))]' + K(t)\Sigma(t)K'(t). \quad \text{(A6)}$$

The matrices $F(\cdot)$ and $H(\cdot)$ are given by

$$F(t, \hat{\gamma}(t-1)) = \frac{\partial}{\partial\gamma} f(t, \gamma(t))\Big|_{\gamma=\hat{\gamma}(t-1|t-1)}, \quad \text{(A7)}$$

$$H(t, \hat{\gamma}(t-1)) = \frac{\partial}{\partial\gamma} h(t, \gamma(t))\Big|_{\gamma=\hat{\gamma}(t-1|t-1)},$$

and

$$P(t|t) = E[\hat{\gamma}(t|t)\hat{\gamma}'(t|t)]. \quad \text{(A8)}$$

REFERENCES

1. P. BLOOMFIELD, "Fourier Analysis of Time Series—An Introduction," Wiley, New York, 1976.

2. R. K. OTNES and L. ENOCHSON, "Applied Time Series," Vol. 1, Wiley, New York, 1978.

3. G. E. P. BOX and G. M. JENKINS, "Time Series Analysis: Forecasting and Control," Holden-Day, San Francisco, California, 1976.

4. H. AKAIKE, *in* "Systems Identification: Advances and Case Studies," p. 27 (R. K. Mehra and D. G. Lainiotis, eds.), Academic Press, New York, 1976.

5. L.-M. LIU, "User's Manual for BMDQ2T (TSPACK): Box-Jenkins Time Series Analysis," *UCLA Biomathematics Department Technical Report*, No. 57, University of California, Los Angeles, 1979.

6. SAS, "SAS/ETS User's Guide," SAS Institute Inc., Cary, North Carolina, 1982.

7. L.-M. LIU, G. B. HUDAK, G. E. P. BOX, M. E. MULLER, and G. C. TIAO, "The SCA System for Univariate Time Series and General Statistical Analysis," Scientific Computing Associates, DeKalb, Illinois, 1983.

8. A. H. HARVEY and G. D. A. PHILLIPS, *Biometrika 66*, 49 (1979)

9. H. AKAIKE, *in* "Direction in Time Series," p. 175 (D. R. Brillinger and G. C. Tiao, eds.), Institute of Mathematical Statistics, Michigan, 1980.

10. D. M. DeLONG, *Proc. Annu. Meet. Am. Stat. Assoc., Stat. Comput. Section, Houston, Texas*, 76 (1980).

11. J. P. INDJEHAGOPIAN, *Cah. Cent. Etud. Rech. Operationelles (Bruxelles) 22*, 397 (1980).

12. V. PANUSKA, *Proc. J. Autom. Control Conf., Ann Arbor, Michigan*, 1014, June 1968.

13. L. H. ZETTERBERG and M. HEROLF, *in* "Quantitative Analysis of the EEG Methods and Applications," *Proc. 2nd Symp. Study Group EEG Methodol.*, p. 461, (M. Matejcek and G. K. Schenk, eds.), Jogny sur Vevey, May, 1975.

14. T. SODERSTRÖM, L. LJUNG, and I. GUSTAVSSON, *Automatica 14*, 231 (1978).

15. P. C. YOUNG and A. JAKEMAN, *Int. J. Control 31*, 741 (1980).

16. P. SHOLL, "The Kalman Filter as an Adaptive ARIMA Model," Ph.D. Dissertation, University of Toledo, Toledo, Ohio, 1982.

17. J. LEDOLTER, *Commun. Stat. A8*, No. 12, 1227 (1979).

18. M. H. SHERIF and L.-M. LIU, *Int. J. Control 40*, 499 (1984).

19. A. H. JAZWINSKI, "Stochastic Processes and Filtering Theory," Academic Press, New York, 1970.

20. B. D. O. ANDERSON and J. B. MOORE, "Optimal Filtering," Prentice-Hall, Englewood Cliffs, New Jersey, 1979.

21. V. PANUSKA, *IEEE Trans. Autom. Control AC-25*, 229 (1980).

22. P. C. YOUNG, *Bull. Inst. Math. Appl. 10*, 209 (1974).

23. S. L. FAGIN, *IEEE Int. Conv. Rec.*, Part I, 216 (1964).

24. K. A. MYERS and B. D. TAPLEY, *IEEE Trans. Atuom. Control AC-15*, 175 (1976).

25. R. K. MEHRA, *IEEE Trans. Autom. Control AC-15*, 175 (1970).

26. D. L. ALSPACH and A. ABIRI, *Symp. Nonlinear Estim. Theory Appl., 3rd, San Diego, California*, 1 (1973).

27. D. L. ALSPACH, *IEEE Trans. Autom. Control AC-19*, 552 (1974).

28. B. ABRAHAM and J. LEDOLTER, "Statistical Methods for Forecasting," Wiley, New York, 1983.

29. J. F. LEATHRUM, *IEEE Trans. Autom. Control AC-26*, 745 (1981).

30. P. NEWBOLD, *Biometrika 61*, 423 (1974).

31. G. M. LJUNG and G. E. P. BOX, *Biometrika 66*, 265 (1979).

32. C. R. SZELAG, *Bell Syst. Techn. J. 61*, 67 (1982).

33. G. C. GOODWIN and R. L. PAYNE, "Dynamic System Identification: Experiment Design and Data Analysis," Academic Press, New York, 1977.

34. L.-M. LIU, *Commun. Stat. B*, in press (1986).

35. L. LJUNG, *IEEE Trans. Autom. Control AC-22*, 551 (1977).

36. L. LJUNG, *IEEE Trans. Autom. Control AC-24*, 36 (1979).

Techniques for Multivariable Self-Tuning Control

H. T. TOIVONEN

Department of Chemical Engineering
Åbo Akademi
SF-20500 Turku (Åbo), Finland

I. INTRODUCTION

Adaptive and self-tuning controllers have been developed as a method for controlling systems whose parameters are unknown and possibly time varying. In this approach on-line identification is used in combination with a controller design method. The procedure is applied to recompute the controller parameters at each sampling time. In practice the method can be used for processes with slowly time-varying parameters in order to keep the controller properly tuned when the process dynamics change. The procedure can also be applied to time-invariant systems when manual tuning is difficult. In this case a self-tuning controller can first be used for tuning the controller parameters and then be removed after the parameters have converged. There is presently a vast literature on adaptive and self-tuning controllers obtained by combining various on-line identification methods and different controller design procedures. Industrial applications of the techniques have also been reported. There are many surveys and books treating various aspects of the field (see, for example, [1-8]).

This chapter deals with methods for self-tuning control of multivariable systems. The study of multivariable self-tuning control is well motivated, as it is difficult to use manual tuning methods for multivariable plants with interacting loops. It also requires time-consuming experiments to determine multivariable process models on which the controller design could be based. A number of procedures for multivariable self-tuning control have been described (see, for example, [9-13]), but few applications of multivariable self-tuning control have been reported.

In this chapter various techniques for multivariable self-tuning control are reviewed. The treatment is restricted to procedures that are designed for controlling stochastic systems. The main control objective is taken to control the system in such a way that the steady-state variances of the inputs and the outputs are as small as possible when the disturbances that affect the system can be described as stochastic processes. An important class of industrial quality control problems can be formulated in this way [14,15]. It is also possible to treat many common design problems, such as reference signal tracking and pole-placement design, in a stochastic framework.

The chapter, which is based on [8], is structured as follows. In Section II a brief description of the general methodology is given. Various techniques for self-tuning control are then described in detail. The procedures are classified into explicit and implicit algorithms [3] depending on how the calculations are organized. In explicit linear quadratic Gaussian (LQG) self-tuning regulators the parameters of an explicit process model are estimated, and the control law is determined by solving a LQG control problem. Various topics in the design of explicit LQG self-tuning regulators are treated in Section III. The implicit algorithms are based on the fact that in some cases there is a close connection between the parameters of the control law and a predictive least squares model. The method of least squares can then be used to estimate the parameters of the predictive model, from which the control law can be determined in a trivial way. In this way the design calculations are reduced significantly. The implicit schemes are classified according to the underlying design method into self-tuning minimum variance controllers, algorithms based on single-step optimal control, and predictor-based procedures for multistep optimal control. Implicit self-tuning controllers for achieving minimum output variance around a reference signal are described in Section IV. Section V deals with self-tuning controllers based on single-step optimal control. A predictor-based procedure for multistep optimal control, which is designed by introducing several predictive least squares models with different prediction times, is described in Section VI.

II. SELF-TUNING CONTROL OF STOCHASTIC SYSTEMS

A. THE CONTROL PROBLEM

Consider a linear discrete-time stochastic system described by the vector difference equation

$$A(q^{-1})y(t) = B(q^{-1})u(t - L - 1) + C(q^{-1})e(t), \tag{1}$$

where u is the p-dimensional input, y is the r-dimensional output, and $\{e(t)\}$ is a Gaussian white noise sequence of prediction errors with zero mean value. In (1), q^{-1} is the backward shift operator [$q^{-1}y(t) = y(t - 1)$, etc.], L represents a time delay, and $A(\cdot)$ $(r \times r)$, $B(\cdot)$ $(r \times p)$, and $C(\cdot)$ $(r \times r)$ are matrix polynomials given by

$$A(z) = I + A_1 z + \cdots + A_n z^n, \tag{2a}$$

$$B(z) = B_0 + B_1 z + \cdots + B_l z^l, \tag{2b}$$

$$C(z) = I + C_1 z + \cdots + C_m z^m. \tag{2c}$$

The parameters of (1) are assumed to be constant or slowly time varying. It is assumed that the zeros of the polynomial det C(z) are outside the closed unit disk. This condition can be considered as fairly mild [7a].

The basic control problem studied here is to control the system described by (1) in such a way that the steady-state variances of the outputs y_i, $i = 1, \ldots, r$, and the inputs u_j, $j = 1, \ldots, p$, are as small as possible. This multiobjective optimization problem is, of course, the basis for linear quadratic Gaussian design [15-17] and the optimal control laws are found by minimizing quadratic loss functions of the form

$$V = \lim_{N\to\infty} E \frac{1}{N} \sum_{t=1}^{N} y(t)^T Q_y y(t) + u(t)^T Q_u u(t), \tag{3}$$

where Q_y and Q_u are positive semidefinite weighting matrices. The controllers obtained by minimizing (3) have the property that it is not possible to reduce any of the closed-loop variances of the outputs y_i or the inputs u_j by changing the control law without increasing the variance of at least one other output or input. The loss function (3) thus gives a convenient parameterization of the optimal control strategies in terms of the weighting matrices Q_y and Q_u. The choice of Q_y and Q_u is made so that a satisfactory combination of closed-loop variances is obtained.

In practice it is often important to include reference signal tracking in the controller design. In the present framework this can be achieved by modifying the quadratic loss function appropriately. The topic will be discussed later in connection with the various algorithms.

B. SELF-TUNING CONTROLLERS

A self-tuning controller is obtained by combining an on-line parameter estimator for estimating the parameters of (1) and a part for designing a control law for the identified model [2]. The adaptive controllers considered here can be described by the following general algorithm.

Algorithm 1: Self-Tuning Controller

Step 1: Parameter Estimation. At time instant t, estimate the parameters of a process model by an on-line identification method based on the measured outputs up to time t.

Step 2: Control Law Computation. Use a control design method to derive a control law for the model obtained in step 1, and determine the corresponding control signal $u^0(t)$.

Step 3. Compute the new signal

$$\bar{u}(t) = u^0(t) + \eta(t), \tag{4}$$

where $\eta(t)$ is an input excitation signal, for example, a PRBS signal or white noise.

Step 4. Determine the input applied to the process at time t as

$$u(t) = \text{sat}(\bar{u}(t);\ \beta,\ \alpha), \tag{5}$$

where $\beta_i \le \alpha_i$ and the function sat(•; •, •) is given componentwise by

$$\mathrm{sat}_i(z;\ \beta,\ \alpha) = \begin{cases} \beta_i, & \text{if } z_i < \beta_i \\ z_i, & \text{if } z_i \in [\beta_i, \alpha_i], \\ \alpha_i, & \text{if } z_i > \alpha_i. \end{cases} \tag{6}$$

Repeat from step 1 at each sampling instant.

The adaptive algorithms have been classified into *explicit* and *implicit* algorithms depending on how the identification in step 1 is organized. In the explicit methods the parameters of (1) are estimated explicitly. In the implicit procedures the system is parameterized in such a way that the controller parameters can be estimated directly or are obtained from the estimated parameters in a trivial way.

In step 2 of the algorithm an optimal control strategy to minimize a quadratic loss function is computed. In order to reduce the computational effort a suboptimal approach is often used in which the loss function (3) is replaced by a finite-time loss function that is applied in a receding-horizon manner. The information that the parameters of the estimated model are uncertain can be exploited in various ways [1]. The most common approach is simply to ignore the parameter uncertainties (certainty equivalence control), but this approach is not optimal [1].

In step 3 the excitation signal $\eta(t)$ is added to the input in order to preserve parameter identifiability in closed-loop operation [18,19]. It can also be used when the parameter estimates are poor in order to obtain better estimates.

Step 4 is introduced for preventing the input from becoming unacceptably large and for taking into account the saturation effects that are always present in practice. It has also been proposed that linear input constraints of the form (6) can be used as design variables, offering a method to reduce control signal variations to an appropriate level [20].

C. PARAMETER ESTIMATION

There are many on-line parameter estimation methods that can be used in self-tuning control algorithms [2,21]. The parameters of (1) can be estimated on-line by a multivariable version of the recursive extended least squares procedure [22] described by the equations

$$\Theta^T(t + 1) = \Theta^T(t) + K(t + 1)\epsilon(t + 1; \Theta(t))^T, \tag{7a}$$

$$K(t + 1) = \frac{P(t)\varphi(t)}{\lambda(t + 1) + \varphi(t)^T P(t)\varphi(t)}, \tag{7b}$$

$$P(t + 1) = \frac{1}{\lambda(t + 1)}\left[P(t) - \frac{P(t)\varphi(t)\varphi(t)^T P(t)}{\lambda(t + 1) + \varphi(t)^T P(t)\varphi(t)}\right], \tag{7c}$$

$$\epsilon(t + 1; \Theta(t)) = y(t + 1) - \Theta(t)\varphi(t), \tag{8}$$

where

$$\Theta(t) = \left[\hat{A}_1(t), \ldots, \hat{A}_n(t), \hat{B}_0(t), \ldots, \hat{B}_l(t), \hat{C}_1(t), \ldots, \hat{C}_m(t)\right] \tag{9}$$

is the estimate of the parameters in (2) and

$$\varphi(t) = [-y(t)^T, \ldots, -y(t + 1 - n)^T, u(t - L)^T, \ldots, u(t - L - l)^T, \epsilon(t; \Theta(t - 1))^T, \ldots, \epsilon(t + 1 - m; \Theta(t - m))^T]^T. \tag{10}$$

There are alternative implementations of the method that may be useful in practical situations [21]. Numerically stable square root algorithms propagate the square root of the matrix P(t) [23,24]. Fast algorithms make use of the fact that the φ vectors have a structure such that $\varphi(t)$ and $\varphi(t + 1)$ are related [21].

A difficulty when estimating the parameters of the system (1) is that both the parameters and the past prediction errors $e(\cdot)$ are unknown and must be estimated simultaneously. In the procedure defined by (7), estimates of the prediction errors are obtained recursively from (8), using the current parameter estimate at each step. A modification of the method is to use the (a posteriori) estimates

$$\epsilon(t;\ \Theta(t)) = y(t) - \Theta(t)\varphi(t - 1) \tag{11}$$

in the regression vector $\varphi(t)$. This modification has been used in order to derive stronger convergence results for a class of adaptive controllers [25].

When $C(q^{-1}) = I$ in (1), the procedure (7) reduces to the recursive least squares method. Then

$$\Theta(t) = \left[\hat{A}_1(t),\ \ldots,\ \hat{A}_n(t),\ \hat{B}_0(t),\ \ldots,\ \hat{B}_1(t)\right], \tag{12a}$$

$$\begin{aligned}\varphi(t) = [&-y(t)^T,\ \ldots,\ -y(t + 1 - n)^T,\\ &u(t - L)^T,\ \ldots,\ u(t - L - 1)^T]^T.\end{aligned} \tag{12b}$$

In this case the recursive algorithm generates the same estimates that could be obtained by using an off-line estimation method for the same input-output data.

In (7), $\lambda(\cdot)$ is an exponential weighting factor that makes it possible to track slowly time-varying parameters. When $\lambda = 1$, all observations are weighted equally, while $\lambda < 1$ gives exponential forgetting of past data. Variable weighting factors, which adapt to changing disturbance levels by retaining a constant amount of information in the estimator, have been developed [26,27] and may be useful in practical situations.

In some applications it is of interest to consider methods based on stochastic approximation, in which the estimates are determined recursively from

$$\Theta^T(t + 1) = \Theta^T(t) + p(t + 1)\varphi(t)\epsilon(t + 1; \Theta(t))^T, \tag{13}$$

where $p(t)$ is a scalar, for example, c/t or $1/\mathrm{tr}\, P(t)^{-1}$ [21]. The modification reduces the computational effort significantly, but the convergence rate of the estimates is in general slower than with the algorithm defined by (7).

D. *STABILITY AND CONVERGENCE*

The closed-loop system obtained when controlling (1) by an adaptive controller is nonlinear, time varying, and stochastic, which makes it difficult to analyze. Some results are available, however, for the case when the parameters of the system (1) are constant.

Goodwin *et al.* [28] apply martingale theory to study an algorithm based on a modified stochastic approximation identification procedure and a minimum variance design. It was shown that subject to a positive realness condition, the inputs and outputs are mean square bounded, and the algorithm converges to the optimal minimum variance controller. For single-input/single-output systems the results have been generalized by Sin and Goodwin [25] to the case when a modified form of the estimation procedure defined by (7) is used.

Ljung [29] has developed a general procedure for analyzing the asymptotic behavior of recursive stochastic algorithms, which can be used to study the adaptive controllers considered here [2,30]. Assuming that $\lambda(t) \to 1$ in the recursive scheme (7), the asymptotic behavior of the adaptive algorithm can be studied in terms of an associated ordinary differential equation. A restriction of this approach is that stability of the closed-loop system must be assumed. The associated ordinary differential equation can be analyzed only in simple cases [29,30]. When it is infeasible to analyze the equation, the results can be used heuristically by solving the associated ordinary differential equation numerically [2,30]. In this way valuable information of the asymptotic behavior of complex adaptive controllers can be gained [31-34].

III. EXPLICIT LQG SELF-TUNING REGULATORS

In this section the design of explicit self-tuning regulators based on linear quadratic Gaussian design is studied. In this approach the parameters of the system (1) are estimated on-line, and the control law is determined by solving a Riccati equation. Algorithms of this form have been described, for example, by Peterka and Astrom [9], Lam [35], Astrom and Zhao-Ying [36], El-Sherief and Sinha [37], and Shieh *et al.* [38].

A. *THE ALGORITHM*

Consider the model

$$\hat{A}(q^{-1})y(t) = \hat{B}(q^{-1})u(t-1) + \hat{C}(q^{-1})\epsilon(t), \tag{14}$$

where a circumflex has been introduced in order to distinguish the model from the true system (1). The model (14) corresponds to (1) with L = 0. This implies no restriction, since the leading matrix coefficients of the B polynomial (2b) can always be set equal to zero. A possible, but nonminimal, state-space representation of (14) is

$$\hat{z}(t+1) = \hat{A}\hat{z}(t) + \hat{B}u(t) + K\epsilon(t+1), \qquad y(t) = [I0 \cdots 0]\hat{z}(t), \tag{15}$$

where

$$\hat{z}(t) = [y(t)^T, \ldots, y(t+1-\hat{n})^T, u(t-1)^T, \ldots, u(t-\hat{l})^T, \epsilon(t)^T, \ldots, \epsilon(t+1-\hat{m})^T]^T \tag{16}$$

and

$$\hat{A} = \begin{bmatrix} F \\ M \end{bmatrix}, \quad \hat{B} = \begin{bmatrix} \hat{B}_0 \\ N \end{bmatrix}, \tag{17}$$

$$F = \left[-\hat{A}_1, \ldots, -\hat{A}_{\hat{n}}, \hat{B}_1, \ldots, \hat{B}_{\hat{l}}, \hat{C}_1, \ldots, \hat{C}_{\hat{m}}\right], \tag{18}$$

and M, N, and K are constant matrices whose structure is evident from the representation (15), (16).

The basic adaptive algorithm based on linear quadratic Gaussian control takes the following form.

Algorithm 2: Explicit LQG Self-Tuning Regulator

Step 1. Determine the parameters of (14) by an on-line parameter estimation method.

Step 2. Compute the control that minimizes the loss function (3) for the estimated model, that is,

$$u^0(t) = -L_t\hat{z}(t), \tag{19}$$

where

$$L_t = \left(\hat{B}^T S_t \hat{B} + Q_u\right)^{-1}\hat{B}^T S_t \hat{A}, \tag{20a}$$

$$S_t = \hat{A}^T S_t \hat{A} - \hat{A}^T S_t \hat{B}\left(\hat{B}^T S_t \hat{B} + Q_u\right)^{-1}\hat{B}^T S_t \hat{A} + Q_x, \tag{20b}$$

$$Q_x = \begin{bmatrix} Q_y & 0 \\ 0 & 0 \end{bmatrix}. \tag{20c}$$

Steps 3 and 4. Proceed as in Algorithm 1.

Repeat from step 1 at each sampling instant.

There are many modifications of the algorithm. One possibility is to use a state-space representation that is of lower order than (15), (16). The state vector will then depend explicitly on the estimated parameters.

When the Riccati equation (20b) is solved iteratively, it is convenient to start the iteration using the Riccati matrix S_{t-1} from the previous sampling period, since it is often a good approximation of the new solution. In order to reduce the computations it has been proposed that only one iteration of the Riccati equation be performed at each sampling instant [35,39, 40]; that is, S_t is determined from

$$S_t = \hat{A}^T S_{t-1} \hat{A} - \hat{A}^T S_{t-1} \hat{B} \left(\hat{B}^T S_{t-1} \hat{B} + Q_u \right)^{-1} \hat{B}^T S_{t-1} \hat{A} + Q_x. \quad (21)$$

This procedure may work well (see, e.g., [8] for a simulated example), but it should be applied with some care. The reason is that as the parameter estimates change from step to step, (21) may in some cases act as a low-pass filter, and therefore the Riccati matrix S_t, and hence the control law (19), may after a period of time be far from the correct solution. This may happen even in cases when the parameter estimates are quite acceptable all the time. This phenomenon has been observed in simulations. It is therefore recommended that the Riccati equation be solved to a given accuracy at each sampling time. More efficient techniques for solving (20b) can, of course, be used, such as doubling algorithms or the matrix sign algorithm [38].

B. SELECTION OF WEIGHTING MATRICES

The weighting matrices Q_y and Q_u in the loss function (3) should be chosen properly to achieve the desired closed-loop performance. The proper values for the weights depend on the system dynamics, and in a truly adaptive procedure the weights should therefore also be adjusted on-line. The adjustment can be done manually for single-input/single-output systems [3,41], but manual tuning may not be well suited for multivariable plants with many inputs and outputs. Therefore it is useful to study methods for tuning the design weights adaptively. One approach is to formulate the control problem as an optimal stochastic control problem with explicit variance restrictions [33,34,42]. In this formulation, the solution is found to the constrained problem

$$\begin{aligned} &\text{Minimize} \quad V = \lim_{N\to\infty} E\,\frac{1}{N}\sum_{t=1}^{N} y(t)^T Q_y y(t), \\ &\text{subject to} \quad \lim_{N\to\infty} E\,\frac{1}{N}\sum_{t=1}^{N} u(t)^T R_i u(t) \le c_1^2, \quad i = 1, \ldots, q, \end{aligned} \tag{22}$$

where the matrices R_i are positive semidefinite. For unstable systems the constraints c_i^2 should be chosen so that the problem has a solution. The formulation (22) has been found useful in industrial applications of stochastic control theory [15,43]. Introducing the vector $\lambda = [\lambda_1, \ldots, \lambda_q]$ of nonnegative Lagrange multipliers, the Lagrangian function of the constrained minimization problem (22) can be written as

$$V_L(\lambda) = \lim_{N\to\infty} E\,\frac{1}{N}\sum_{t=1}^{N} y(t)^T Q_y y(t) + u(t)^T Q_u(\lambda) u(t), \tag{23}$$

where

$$Q_u(\lambda) = \sum_{i=1}^{q} \lambda_i R_i. \tag{24}$$

The solution of the constrained problem (22) is conveniently found by solving an associated dual maximization problem [44]. By applying an on-line gradient method for solving the dual problem a two-phase adaptive algorithm is obtained as follows [33,34,42]. (1) Apply a self-tuning controller based on linear quadratic Gaussian design for minimizing the loss function $V_L(\lambda)$, (23), with the input weighting matrix given by (24) (Algorithm 2). (2) At each sampling instant, adjust the Lagrange multipliers by the stochastic approximation scheme

$$\lambda_i(t + 1) = \lambda_i(t) + \mu_t \lambda_i(t)\left(u(t)^T R_i u(t) - c_i^2\right)\Big/c_i^2, \qquad i = 1, \ldots, q, \tag{25}$$

where μ_t is a small positive scalar.

C. USE OF LEAST SQUARES MODELS

The least squares method is a convenient procedure for parameter estimation, and it is often used in practice. It is therefore worth studying to what extent it is possible to base the self-tuning controller described by Algorithm 2 on a least squares model

$$\hat{A}(q^{-1})y(t) = \hat{B}(q^{-1})u(t - 1) + \epsilon(t) \tag{26}$$

in cases when the system is correctly described by (14) with $\hat{C}(q^{-1}) \neq I$. One motivation for using the model (26) is that it can be considered as an approximation to (14) obtained by multiplying (14) by $\hat{C}(q^{-1})^{-1}$ and approximating the inverse by a matrix polynomial of finite order. In general this approach has the drawback that high orders for $\hat{A}(\cdot)$ and $\hat{B}(\cdot)$ may have to be used in (26) in order to obtain a good approximation. For adaptive algorithms in closed-loop operation, there are, however, some experimental results that indicate that quite good performance may be obtained even with low-order least squares models of the form (26) [31,33,34,42]. Intuitively this can be understood as follows. When a system described by (14) with the state-space representation (15) is controlled by a time-invariant feedback of the form (19), the state vector $\hat{z}(t)$ can be constructed from the inputs u and the outputs y only using (15) and (19) and the relation

$$\epsilon(t + 1) = y(t + 1) - F\hat{z}(t) - \hat{B}_0 u(t)$$

for eliminating the prediction errors. It follows that the output of the closed-loop system is correctly predicted by a least squares model of the form (26), which holds for the given closed-loop conditions. The order of the least squares model is given by the observability index of the pair $(L, \hat{A} - KF)$. It follows that an adaptive controller that is based on a least squares model may in some cases converge to the required strategy. One example is when the self-tuning controller is based on a single-step loss function [12,13]. Numerical examples have shown that good results can also be obtained when the design is based on a multistep loss function. Several numerical examples are presented in [31] for minimum variance control of nonminimum phase systems and in [33,34,42] for the variance-constrained optimal control problem defined by (22). In these references self-tuning controllers are based on the least squares model (26) and applied to systems described by (14). It was found that in many circumstances the adaptive algorithm converges to a near-optimal control law, giving a value of the loss that is only a few percent larger than the minimum loss obtained when using the optimal strategy. In many examples the nonoptimality of the convergence points could not be discovered by simulations, but was determined by studying the associated ordinary differential equation of Ljung [29,30]. In view of these empirical results it is clearly an interesting alternative to use the least squares model (26), although the behavior of the algorithms is not well understood in situations when the model and the system are not compatible.

D. *CAUTIOUS CONTROL*

Algorithm 2 is a certainty equivalence controller [1]; that is, the control signal is determined under the assumption that the estimated parameters obtained in step 1 are the correct ones. When the parameter estimates are poor, the approach may not work well because too much confidence is put on the current estimates. It could then be useful to take the uncertainty of the parameter estimates into account in some way. A fairly simple suboptimal method for achieving this is offered by the approach known as open-loop optimal feedback control or cautious control [1]. In this approach the loss function

$$V_N = E\left[\sum_{i=t}^{t+N} y(i)^T Q_y y(i) + u(i)^T Q_u u(i) \,\middle|\, y(t),\right.$$
$$\left. u(t-1),\ y(t-1),\ \ldots\right], \tag{27}$$

where $E[\cdot|\cdot]$ denotes conditional expectation, is minimized at each step, taking the uncertainty of the current parameter estimates into account, but under the assumption that future measurements are not used for improving the estimates. The first control $u^0(t)$ in the optimal control sequence is then applied, and the procedure is repeated at each sampling time. The approach is suboptimal, since the fact that the control will influence future estimates and their accuracy is not taken into account. The effect of parameter uncertainties in this method is simply to make the controller more cautious, since less confidence is put on the parameter estimates that are available at each sampling instant.

Peterka and Åström [9] design a cautious controller by modeling the system parameters as random white noise processes. The algorithm described in [9] is based on a square-root implementation of the least squares method for parameter estimation. Here we describe the algorithm which is obtained when using the recursive least squares method defined by (7) and (12) [33]. In this approach the parameter uncertainties can be described as follows. Assume that the recursive least squares method is used to estimate the parameters of a system that is described by the least squares model (26). Let S be a known matrix of appropriate dimension. The covariances of the estimated parameters $\Theta(t)$ are then described by the equation [24,33]

$$E\left[(\Theta(t) - \Theta_0)^T S(\Theta(t) - \Theta_0)\right] = P(t) \text{ tr } SR_\epsilon, \tag{28}$$

where Θ_0 is the true parameter matrix, $P(t)$ is the matrix defined by (7), and R_ϵ is the covariance matrix of the white noise ϵ of (26). In practice, R_ϵ is not known and it must be estimated. The maximum likelihood estimate of R_ϵ can be determined recursively from [24]

$$\gamma(t+1)\hat{R}_\epsilon(t+1) = \lambda(t+1) \times \left[r(t)\hat{R}_\epsilon(t) + \frac{\epsilon(t+1;\,\Theta(t))\,\epsilon(t+1;\,\Theta(t))^T}{\lambda(t+1) + \varphi(t)^T P(t)\varphi(t)}\right] \tag{29a}$$

$$\gamma(t+1) = \lambda(t)\gamma(t) + 1. \tag{29b}$$

These relations can now be used to obtain the following algorithm for cautious control [33].

Algorithm 3: Cautious Control

Step 1. Estimate the parameters of (26) by the recursive least squares method using (7) and (29) to obtain $\Theta(t)$, $P(t)$, and $\hat{R}_\epsilon(t)$. Here it is convenient to reorder the parameter matrix and to use

$$\Theta(t) = \left[\hat{B}_0, \hat{A}_1, \ldots, \hat{A}_{\hat{n}}, \hat{B}_1, \ldots, \hat{B}_{\hat{l}}\right],$$

$$\varphi(t) = [u(t)^T, -y(t)^T, \ldots, -y(t+1-\hat{n})^T, \qquad (30)$$
$$u(t-1)^T, \ldots, u(t-\hat{l})^T]^T$$

instead of (12).

Step 2. Use the estimates and their uncertainties to find the control sequence that minimizes the loss function (27) under the assumption that future measurements are not used for improving the parameter estimates. This problem can be approached by solving an optimal control problem for a system with random parameters. The optimal control sequence can be found by dynamic programming [45]. Introducing the state space representation defined by Eqs. (15) through (18) and using the result (28) to describe the parameter uncertainties gives the solution [33]

$$u^0(t) = -L(t)z(t), \qquad (31)$$

where

$$L(k) = \left[\hat{B}^T S(k+1)\hat{B} + P_{11}(t)\ \mathrm{tr}\left(S_{11}(k+1)\hat{R}_\epsilon(t)\right) + Q_u\right]^{-1}$$
$$\times \left[\hat{B}^T S(k+1)\hat{A} + P_{12}(t)\ \mathrm{tr}\left(S_{11}(k+1)\hat{R}_\epsilon(t)\right)\right],$$

$$S(k) = \hat{A}^T S(k+1)\hat{A} + P_{22}(t)\ \operatorname{tr}\left(S_{11}(k+1)\hat{R}_\epsilon(t)\right)$$

$$- \left[\hat{A}^T S(k+1)\hat{B} + P_{12}^T(t)\ \operatorname{tr}\left(S_{11}(k+1)\hat{R}_\epsilon(t)\right)\right] L(k)$$

$$+ Q_x,$$

$$S(t+N) = Q_x, \quad k = t+N-1, \ldots, t. \tag{32}$$

Here Q_x is defined by (20c). In (32), the Riccati matrix S is partitioned as

$$S = \begin{bmatrix} S_{11} & S_{12} \\ S_{12}^T & S_{22} \end{bmatrix}, \quad \text{(first block row and column of size } r\text{)}$$

and the matrix P(t) of (7) as [cf. (30)]

$$P(t) = \begin{bmatrix} P_{11}(t) & P_{12}(t) \\ P_{12}^T(t) & P_{22}(t) \end{bmatrix}. \quad \text{(first block row and column of size } p\text{)}$$

Steps 3 and 4. Proceed as in Algorithm 1.

Repeat from step 1 at each sampling instant.

A possible difficulty in cautious control is the so-called turn-off phenomenon [1]. This means that when the estimates are poor, the parameter uncertainties will cause the control action to be cautious, and the control signal may become very small. Because the control does not excite the system, the estimates may become still worse. In this way the control may be turned off for some period of time, until the system is excited in such a way that better estimates are obtained. One way of avoiding this difficulty in Algorithm 3 is to apply the input excitation signal according to (4) when the estimates are poor.

IV. SELF-TUNING MINIMUM VARIANCE CONTROLLERS

This section is concerned with the case when the criterion for control is to achieve minimum variance of the outputs without constraints on the input variances, that is, when $Q_u = 0$ in the loss function (3). In this case there is a class of systems for which it is possible to parameterize the model in such a way that the design calculations are trivial. It is also straightforward to include reference signal tracking and feedforward from measured signals in this approach. The basic self-tuning minimum variance regulator of this kind was given for single-input/single-output systems by Åström and Wittenmark [46], and a multivariable generalization has been given by Borisson [10,11].

A. *THE CONTROL PROBLEM AND A SELF-TUNING CONTROLLER*

Consider a stochastic system described by

$$A(q^{-1})y(t) = B(q^{-1})u(t - L - 1) + D(q^{-1})v(t - L - 1) + C(q^{-1})e(t), \tag{33}$$

where v(t) is a known signal. It is assumed that the number r of outputs does not exceed the number p of inputs, that rank $B(0) = r$, and that the system is stably invertible [7a].

Now the loss function is defined as

$$V_{MV} = \lim_{N\to\infty} E\,\frac{1}{N}\sum_{t=1}^{N} [y(t + L + 1) - y_r(t + L + 1)]^T \times Q_y[y(t + L + 1) - y_r(t + L + 1)], \tag{34}$$

where $y_r(\cdot)$ is a reference signal to be tracked and assumed known at least L + 1 steps ahead. The control strategy that minimizes (34) is found by considering the model that predicts the output of the system (33) L + 1 steps ahead and that is given by [10,11]

$$y(t + L + 1) = \tilde{C}(q^{-1})^{-1}[\tilde{G}(q^{-1})y(t) + \tilde{F}(q^{-1})B(q^{-1})u(t) + \tilde{F}(q^{-1})D(q^{-1})v(t)] + F(q^{-1})e(t + L + 1), \tag{35}$$

where the matrix polynomials $\tilde{C}(\cdot)$, $\tilde{G}(\cdot)$, $\tilde{F}(\cdot)$, and $F(\cdot)$ are defined by the identities

$$\begin{aligned} C(z) &= A(z)F(z) + z^{L+1}G(z), \quad \deg F(z) = L, \\ \tilde{F}(z)G(z) &= \tilde{G}(z)F(z), \\ \det \tilde{F}(z) &= \det F(z), \quad \tilde{F}(0) = F(0) = I, \\ \tilde{C}(z) &= \tilde{F}(z)A(z) + z^{L+1}\tilde{G}(z), \quad \det \tilde{C}(z) = \det C(z). \end{aligned} \tag{36}$$

Since the term $F(q^{-1})e(t + L + 1)$ is uncorrelated with $y(t)$, $y(t - 1)$, ..., $u(t - 1)$, $u(t - 2)$, ..., $v(t)$, $v(t - 1)$, ..., the strategy that minimizes (34) is obtained by setting the first expression on the right-hand side of (35) equal to $y_r(t + L + 1)$. Using the assumption that $\det C(z)$, and hence $\det \tilde{C}(z)$, has all zeros outside the closed unit disk, the resulting strategy is in steady state given by the equation

$$\tilde{F}(q^{-1})B(q^{-1})u(t) + \tilde{G}(q^{-1})y(t) + \tilde{F}(q^{-1})D(q^{-1})v(t) - \tilde{C}(q^{-1})y_r(t + L + 1) = 0. \tag{37}$$

The condition that the system is stably invertible is necessary for the strategy defined by (37) to be realizable with a bounded input sequence (cf. [7a]). The output of the closed-loop system is in steady state given by

$$y(t + L + 1) = y_r(t + L + 1) + F(q^{-1})e(t + L + 1). \tag{38}$$

The strategy gives minimum variance of the outputs and is independent of the weight Q_y. The minimum variance strategy (37) can be written compactly as

$$B_0 u(t) + \Theta\varphi(t) = y_r(t + L + 1), \tag{39}$$

where $B_0 = B(0)$,

$$\varphi(t) = \Big[y(t)^T, \ldots, y(t - n_y)^T, u(t - 1)^T, \ldots, u(t - n_u)^T, v(t)^T, \ldots, v(t - n_v)^T, y_r(t + L)^T, \ldots, y_r(t + L - n_{y_r})^T\Big]^T,$$

and Θ consists of the matrix coefficients of the polynomials $\tilde{G}(z)$, $\tilde{F}(z)B(z) - B(0)$, $\tilde{F}(z)D(z)$, and $I - \tilde{C}(z)$.

When the number of inputs and the number of outputs are equal, (37), or (39), defines the minimum variance control law uniquely,

$$u_{MV}(t) = -B_0^{-1}[\Theta\varphi(t) - y_r(t + L + 1)]. \tag{40}$$

When the number of inputs exceeds the number of outputs, the minimum variance strategy is not unique. The nonuniqueness can then be exploited, for example, by selecting the control signal so as to give minimum output variance with least control energy at each step [47]. The control signal is then found by solving

$$\min u(t)^T P u(t) \quad \text{subject to (39)}, \tag{41}$$

where P is a strictly positive definite matrix. The solution to (41) is

$$u_{MV}(t) = -P^{-1}B_0^T\left(B_0 P^{-1} B_0^T\right)^{-1}[\Theta\varphi(t) - y_r(t + L + 1)]. \tag{42}$$

An alternative is to determine the input so as to give minimum output variance with the smallest changes of the control signal at each step, that is, by solving

$$\min[u(t) - u(t-1)]^T P[u(t) - u(t-1)] \quad \text{subject to (39)}. \tag{43}$$

The resulting control law is

$$u_{MV}(t) = u(t-1) - P^{-1}B_0^T\left(B_0P^{-1}B_0^T\right)^{-1} [\Theta\varphi(t) - y_r(t+L+1) + B_0u(t-1)]. \tag{44}$$

The structure of the minimum variance control law suggests that a self-tuning minimum variance controller could be designed by estimating the parameters of a predictive model corresponding to (35) and setting the predicted output equal to the reference value $y_r(t + L + 1)$ at each sampling instant [46,10,11]. In this way the following simple implicit algorithm is obtained.

Algorithm 4: Minimum Variance Control
(dim u $\geq$ dim y)

Step 1. Estimate the parameters of the predictive model

$$y(t+L+1) = \mathscr{A}(q^{-1})y(t) + \mathscr{B}(q^{-1})u(t) + \mathscr{C}(q^{-1})y_r(t+L) + \mathscr{D}(q^{-1})v(t) + \epsilon(t+L+1) \tag{45}$$

by the method of least squares. Here L is the time delay of the system (33), which is assumed to be known.

Step 2. Determine the control signal $u_{MV}(t)$ from the minimum variance strategy

$$\mathscr{A}(q^{-1})y(t) + \mathscr{B}(q^{-1})u(t) + \mathscr{C}(q^{-1})y_r(t+L) + \mathscr{D}(q^{-1})v(t) = y_r(t+L+1) \tag{46}$$

using (40), (42), or (44), depending on the situation.

Repeat from step 1 at each sampling instant.

Note that the reference signal $y_r(t)$ can be determined as the output from a reference model,

$$A_r(q^{-1})y_r(t + L + 1) = B_r(q^{-1})u_r(t),$$

and in this way the algorithm is closely connected with model reference adaptive control systems [2,4].

Algorithm 4 and various versions of it have been studied extensively and a number of results concerning stability and convergence are available. Goodwin *et al.* [28] have applied the Martingale theory to establish stability and convergence of a version of Algorithm 4, in which a modified stochastic approximation procedure is used to estimate the parameters. It is shown that provided a certain transfer function associated with the system (33) is positive real, the inputs and outputs are mean square bounded, and the algorithm converges to the minimum variance controller. For single-input/single-output systems the result has been generalized to the case when a slightly modified version of the extended least squares method (7) is used for estimation [25].

If it assumed that the signals remain bounded, the associated ordinary differential equation of Ljung [29,30] can be used to study the asymptotic behavior of the algorithm. The differential equation associated with Algorithm 4 can be analyzed by applying the procedure described in [22]. The result shows that the algorithm converges to the minimum variance strategy (37) if the transfer function

$$\tilde{C}(q^{-1})^{-1} - \frac{1}{2}I$$

is strictly positive real, that is, if the matrix

$$\mathrm{Re}[\tilde{C}(e^{i\omega})^{-1} + \tilde{C}(e^{i\omega})^{-T}] - I$$

is strictly positive definite for all ω.

B. SYSTEMS WITH ARBITRARY TIME DELAYS

Algorithm 4 is restricted to the case when the matrix $B_0 = B(0)$ has full rank. This implies a restriction on the time delays of the system. It is, however, straightforward to generalize the approach to systems with arbitrary but known time delays. Dugard *et al.* [48] give a procedure that makes use of the system interactor matrix, and a slight modification of the method is now described. For the present treatment it is convenient to consider the system (1) written in the transfer function form

$$y(t) = q^{-L-1}H(q^{-1})u(t) + N(q^{-1})e(t), \tag{47}$$

where

$$H(q^{-1}) = \left[q^{-L_{ij}} \frac{b_{ij}(q^{-1})}{a_{ij}(q^{-1})} \right],$$

$$a_{ij}(0) = 1, \quad b_{ij}(0) \neq 0, \quad \min_{i,j} L_{ij} = 0, \tag{48}$$

Then there exist diagonal matrices

$$D_1(q) = \mathrm{diag}(q^{k_1}, \ldots, q^{k_r}), \quad \min_i k_i = 0, \tag{49}$$

$$D_2(q) = \mathrm{diag}(q^{l_1}, \ldots, q^{l_p}), \quad \min_i l_i = 0, \tag{50}$$

such that the matrix

$$K = \lim_{q^{-1}\to 0} D_1(q)H(q^{-1})D_2(q) \tag{51}$$

is finite and has no zero rows and at least r nonzero columns. It is assumed that the matrix K formed in this way has full rank. Introduce the transfer function

$$\bar{H}(q^{-1}) = D_1(q)H(q^{-1})D_2(q) \tag{52}$$

and the signals

$$\bar{y}(t) = D_1(q)y(t), \tag{53}$$

$$\bar{u}(t) = q^{l_{max}} D_2(q^{-1})u(t), \quad l_{max} = \max_i l_i. \tag{54}$$

From (47) we then have

$$\bar{y}(t) = q^{-\bar{L}-1}\bar{H}(q^{-1})\bar{u}(t) + D_1(q^{-1})N(q^{-1})e(t), \tag{55}$$

where $\bar{L} = L + l_{max}$. From (55) a predictive model analogous to (35) can be derived for the signal $\bar{y}(t)$, giving a model of the form [48]

$$\bar{y}(t + \bar{L} + 1) = \bar{C}(q^{-1})^{-1}[\bar{\mathscr{A}}(q^{-1})y(t) + \bar{\mathscr{B}}(q^{-1})\bar{u}(t)]$$
$$+ \bar{F}(q^{-1})e(t + \bar{L} + 1), \tag{56}$$

where $\bar{\mathscr{B}}(0) = K$. Setting the predicted value of the signal $\bar{y}(\cdot)$ equal to $D_1(q)y_r(t + \bar{L} + 1)$ now gives a minimum variance strategy corresponding to (37),

$$\bar{\mathscr{B}}(q^{-1})\bar{u}(t) + \bar{\mathscr{A}}(q^{-1})y(t) - \bar{C}(q^{-1})D_1(q)y_r(t + \bar{L} + 1) = 0. \tag{57}$$

From (56) and (57) it follows that Algorithm 4 can be generalized to systems with arbitrary but known time delays by basing the algorithm on the model

$$\bar{y}(t + \bar{L} + 1) = \mathscr{A}(q^{-1})y(t) + \mathscr{B}(q^{-1})\bar{u}(t)$$
$$+ \mathscr{C}(q^{-1})\left(D_1(q)y_r(t + \bar{L})\right) + \epsilon(t + \bar{L} + 1) \tag{58}$$

and applying the control strategy

$$\mathscr{A}(q^{-1})y(t) + \mathscr{B}(q^{-1})\bar{u}(t) + \mathscr{C}(q^{-1})\left(D_1(q)y_r(t + \bar{L})\right)$$
$$= D_1(q)y_r(t + \bar{L} + 1), \tag{59}$$

where the feedforward signal has been left out for convenience.

When there are different delays at the outputs only, that is, when $D_2(q) = I$, (56) corresponds to using different prediction times for the various outputs, and (57) defines the optimal strategy that minimizes (34). In [20] a numerical example is shown for this case in which a self-tuning controller based on the model (58) is applied. When there are different delays at the inputs, corresponding to a nondiagonal system interactor matrix [48], (54) corresponds to introducing additional time delays, since for the ith input we have $\bar{u}_i(t) = u_i(t + l_{max} - l_i)$; that is, the signal $\bar{u}_i(t)$ is determined at time instant t and applied to the process at the later time $t + l_{max} - \ell_i$. In this case the approach is suboptimal, since the fact that some inputs affect the outputs with a delay less than $\bar{L}$ is not exploited. Dugard *et al.* [48] give a suboptimal procedure that is closely related to the method described here and that has the feature that $\bar{L} + 1$ control laws are applied cyclically.

V. SELF-TUNING CONTROLLERS BASED ON SINGLE-STEP OPTIMAL CONTROL

A. *THE CONTROL PROBLEM AND A SELF-TUNING CONTROLLER*

Minimum variance control has the property that the strategy may generate large inputs [17]. In order to achieve optimal control with reduced input variances, linear quadratic Gaussian design can be used, but more complex algorithms are then required for self-tuning control. A suboptimal approach, which preserves the computational simplicity of the method based on minimum variance control, is to base the design on the single-step loss function [49,50,12,13]

$$V_1 = E\Big[\|y(t + L + 1) - y_r(t + L + 1)\|^2_{Q_y}$$
$$+ \|P(q^{-1})u(t)\|^2_{Q_u} \mid y(t),\ u(t),\ v(t),\ y(t - 1),\ \ldots\Big], \tag{60}$$

where $\|x\|^2_Q = x^TQx$ and $P(q^{-1})$ is a (matrix) polynomial.

The strategy obtained by minimizing (60) is not optimal for the problem described in Section II,A, which involves the steady-state input and output variances. It has been shown, however, that good steady-state performance in many cases can be obtained by basing the design on a single-step loss function [51]. Single-step optimal control then provides a simple suboptimal method for reducing the input variances.

Now consider a system described by (33). In view of the predictive model (35) the control law that minimizes (60) is defined by the equation [12,13]

$$B_0^TQ_y\tilde{C}(q^{-1})^{-1}[\tilde{G}(q^{-1})y(t) + \tilde{F}(q^{-1})B(q^{-1})u(t)$$
$$+ \tilde{F}(q^{-1})D(q^{-1})v(t)] + P(0)^TQ_uP(q^{-1})u(t)$$
$$= B_0^TQ_yy_r(t + L + 1). \tag{61}$$

Self-tuning controllers based on single-step optimal control have been designed by Clarke and Gawthrop [49,50] for single-input/single-output systems, and the procedure has been generalized to a class of multivariable systems by Koivo [12] and Keviczky and Kumar [13]. The approach in these papers is based on the fact that when there is an equal number of inputs and outputs, it follows from (35) and (61) that the control law defined by (61) is equivalent with the strategy that minimizes the steady-state variance of the signal

$$\phi(t + L + 1) = y(t + L + 1) + B_0\left(B_0^T Q_y B_0\right)^{-1} P(0)^T Q_u P(q^{-1}) u(t). \qquad (62)$$

Hence a self-tuning controller for minimizing (60) is obtained by applying Algorithm 4 to achieve minimum variance control of the signal ϕ, using the predictive model

$$\phi(t + L + 1) = \mathscr{A}(q^{-1}) y(t) + \mathscr{B}(q^{-1}) u(t) + \mathscr{C}(q^{-1}) y_r(t + L) + \mathscr{D}(q^{-1}) v(t) + \epsilon(t + L + 1) \qquad (63)$$

and the control law (46). This approach is restricted to systems with an equal number of inputs and outputs. It also has the drawback that there is no convenient way of estimating the matrix B_0 in the algorithm, since this parameter appears nonlinearly in (62). Therefore B_0 is usually assumed to be known [12,13].

In order to design an adaptive algorithm that does not have these restrictions, a more direct approach similar to Algorithm 4 could be attempted by using the predictive model (45) in combination with a control law for minimizing (60) [52]. To study the properties of an algorithm designed in this way, it is useful to observe that in the general case with no restrictions on the number of inputs and outputs the strategy defined by (61) can in steady state be written in the form [53]

$$B_0^T Q_y [\mathscr{A}^0(q^{-1}) y(t) + \mathscr{B}^0(q^{-1}) u(t) + \mathscr{D}^0(q^{-1}) v(t)] + c(q^{-1}) P(0)^T Q_u P(q^{-1}) u(t) - B_0^T Q_y c(q^{-1}) y_r(t + L + 1) = 0, \qquad (64)$$

where

$$c(q^{-1}) = \det \tilde{C}(q^{-1}) = \det C(q^{-1}),$$

$$\mathscr{A}^0(q^{-1}) = \operatorname{adj}[\tilde{C}(q^{-1})]\tilde{G}(q^{-1}),$$

$$\mathscr{B}^0(q^{-1}) = \operatorname{adj}[\tilde{C}(q^{-1})]\tilde{F}(q^{-1})B(q^{-1}), \quad \mathscr{B}^0(0) = B(0) = B_0,$$

$$\mathscr{D}^0(q^{-1}) = \operatorname{adj}[\tilde{C}(q^{-1})]D(q^{-1}).$$

An implicit self-tuning controller analogous with Algorithm 4 is now obtained as follows [53].

Algorithm 5: Single-Step Optimal Control
(dim y, dim u arbitrary)

Step 1. Estimate the parameters of the predictive model (45) by the method of least squares.

Step 2. Determine the control signal so as to minimize the loss function (60) using the model obtained in step 1; that is, determine u(t) from

$$\mathscr{B}_0^T Q_y\left[\mathscr{A}(q^{-1})y(t) + \mathscr{B}(q^{-1})u(t) + \mathscr{C}(q^{-1})y_r(t+L) + \mathscr{D}(q^{-1})v(t)\right]$$

$$+ P(0)^T Q_u P(q^{-1})u(t) = \mathscr{B}_0^T Q_y y_r(t+L+1), \tag{65}$$

where $\mathscr{B}_0 = \mathscr{B}(0)$.

Repeat from step 1 at each sampling instant.

In this algorithm there are no restrictions on the number of inputs and outputs. Systems with arbitrary time delays can be treated by the procedure described in Section IV,B.

Note that in the required strategy (64) the input weight parameters Q_u and $P(q^{-1})$ and the system-dependent polynomial $c(q^{-1})$ are coupled in a way that is ignored in the implicit scheme described by Algorithm 5. From (64) and (65) it follows that convergence of the algorithm to the required strategy implies that the estimated matrix polynomial $\mathscr{B}(q^{-1})$ should converge to some value that depends on the input weight parameters as well as $c(q^{-1})$. The convergence behavior is hence more involved than that of Algorithm 4. The convergence of Algorithm 5 has been studied in [53] using the approach based on the associated ordinary differential equation [29,30]. It was shown for the case when the parameter B_0 is known that convergence to the required strategy (64) is actually obtained if the model order is chosen so that (64) and (65) are compatible and if the transfer function $[1/c(q^{-1})] - \frac{1}{2}$ is strictly positive real.

B. *GENERALIZED COST FUNCTIONS*

The single-step cost function (60) can be generalized according to [13] as

$$V_1 = E\Big[\|W_y(q^{-1})y(t + L + 1) - W_r(q^{-1})y_r(t + L + 1)\|^2_{Q_y} + \|W_u(q^{-1})u(t)\|^2_{Q_u} \mid y(t), u(t), v(t), y(t - 1), \ldots\Big], \tag{66}$$

where $W_y(\cdot)$, $W_r(\cdot)$, and $W_u(\cdot)$ are matrix fraction descriptions. This case can be treated as follows. Let $W_u(\cdot)$ be expressed by the right prime matrix fraction description

$$W_u(q^{-1}) = P(q^{-1})W_{ud}(q^{-1})^{-1}$$

and introduce the filtered signals

$$\begin{aligned} y_F(t) &= W_y(q^{-1})y(t), \\ y_{r,F}(t) &= W_r(q^{-1})y_r(t), \\ u_F(t) &= W_{ud}(q^{-1})^{-1}u(t). \end{aligned} \tag{67}$$

The loss function (66) then takes the form (60) in terms of the signals (67). From (33) a predictive model similar to (35) can be derived for the filtered signals [13]. It follows that a self-tuning controller for minimizing (66) is obtained by applying Algorithm 5 to the filtered signals (67).

The design weights in (66) can be chosen to affect the closed-loop response. For stably invertible systems with an equal number of inputs and outputs, one possibility is to set $W_u(q^{-1}) = 0$, which results in the closed-loop response [cf. (38) and (67)]

$$W_y(q^{-1})y(t) = W_r(q^{-1})y_r(t) + \epsilon(t). \tag{68}$$

VI. PREDICTOR-BASED PROCEDURES FOR MULTISTEP OPTIMAL CONTROL

When the controller design is based on the single-step loss functions (60) or (66), it is possible to construct the simple implicit self-tuning controller described by Algorithm 5. The procedure is, however, not optimal for the stochastic control problem described in Section II,A, and it is not well suited for systems with unknown time delays or other nonminimum phase properties. Better control performance can in general be obtained by minimizing a multistep loss function. Menga and Mosca [54] have given a procedure that can be regarded as a generalization of Algorithm 5 to the case when a multistep loss function is to be minimized. In the approach several predictive models with different prediction times are used for predicting the outputs and the inputs at future time instants.

If a constant linear feedback law is applied to the system described by (1) at time instants $t + 1$, $t + 2$, ..., the outputs $y(t + 1)$, ..., $y(t + N)$ and the inputs $u(t + 1)$, ..., $u(t + N)$ can be predicted at the time instant t. Menga and

Mosca [54] introduce the predictive models

$$y(t + i) = \mu_i u(t) + \Theta_i \varphi(t) + w_i(t + i),$$
$$u(t + i) = \phi_i u(t) + \psi_i \varphi(t) + v_i(t + i), \quad i = 1, \ldots, N, \tag{69}$$

where $\varphi(t)$ is an appropriately chosen vector consisting of past inputs and outputs and $w_i(t + i)$, $v_i(t + i)$, $i = 1, \ldots, N$, are random vectors uncorrelated with $u(t)$ and $\varphi(t)$. Minimizing the N-step loss function (27) using the predictive models (69) gives the control law

$$u(t) = -\left[Q_u + \sum_{i=1}^{N} \left(\mu_i^T Q_y \mu_i + \phi_i^T Q_u \phi_i\right)\right]^{-1} \times \sum_{i=1}^{N} \left(\mu_i^T Q_y \theta_i + \phi_i^T Q_u \psi_i\right) \varphi(t). \tag{70}$$

An adaptive controller similar to Algorithm 5 can now be designed by estimating the parameters of the predictive models (69) using the method of least squares and by applying the control law (70). Simulation studies have shown that the procedure has nice properties, and it can be applied, for example, to systems with unknown time delays and systems that are not stably invertible [55]. It is straightforward to include reference signal tracking in the method [55]. Some theoretical results for the procedure have been described [55]. The algorithm is, however, difficult to analyze because the parameters of the predictive equations (69) will be nonlinear functions of the parameters of the feedback law.

VII. SUMMARY AND CONCLUSIONS

Techniques for self-tuning control of multivariable stochastic systems have been described. The basic control problem that has been considered is to minimize the steady-state variances of the inputs and the outputs. Both explicit LQG self-tuning regulators and various implicit adaptive algorithms have been treated.

The optimal solution of the stochastic control problem leads to linear quadratic Gaussian design. The explicit LQG self-tuning regulators involve the on-line solution of a steady-state Riccati equation. This increases the computational requirements of the algorithm, but instead the approach has the advantage that, in contrast to most implicit schemes, it can be applied to systems with unknown time delays and systems that are not stably invertible. It has been shown that in this approach it is straightforward to take parameter uncertainties into account in a suboptimal manner by an open-loop optimal feedback-type procedure.

The implicit self-tuning minimum variance controller is designed for the case when the criterion for control is to achieve minimum output variance without constraints on the input variances. The algorithm is appealing due to its simplicity, consisting of a predictive least squares model and a control law that sets the predicted output equal to the reference value. A major restriction of the procedure is the condition that the system should be stably invertible [7a]. It has been shown how arbitrary time delays can be handled in this approach if they are known. In order to achieve minimum variance control of systems with unknown time delays and systems that are not stably invertible, the more complex explicit algorithm based on the solution of a Riccati equation should be used.

The implicit adaptive algorithms based on single-step optimal control are generalizations of the self-tuning minimum variance controller, preserving its computational simplicity, but with design parameters by which the closed-loop behavior can be affected. The approach can, for example, be applied to systems that are not stably invertible, by selecting the design weights of a generalized single-step cost function properly. It is, however, not always straightforward to find proper values for the design parameters in such cases. The design method is also suboptimal since a single-step loss function is minimized.

A useful generalization of the implicit predictor-based procedures is obtained by introducing several predictive least squares models with different prediction times. The control signal is then determined to minimize a multistep loss function based on predicted outputs and inputs. Since a multistep loss function is minimized, the procedure is useful for systems that are not stably invertible.

Many of the adaptive techniques that have been described have been developed only recently, and, therefore, very few applications of the methods have been reported. In recent years there has been an increasing interest in industrial applications of self-tuning controllers designed for single-input/single-output systems [2,3,5,6], however, and the multivariable self-tuning controllers seem promising for practical applications as well.

REFERENCES

1. B. WITTENMARK, *Int. J. Control 21*, 705 (1975).
2. K. J. ÅSTRÖM, U. BORISSON, L. LJUNG, and B. WITTENMARK, *Automatica 13*, 457 (1977).
3. K. J. ÅSTRÖM, *Automatica 19*, 471 (1983).

4. I. D. LANDAU, "Adaptive Control—The Model Reference Approach," Dekker, New York, 1979.

5. K. S. NARENDRA and R. V. MONOPOLI (ed.), "Applications of Adaptive Control," Academic Press, New York, 1980.

6. H. UNBEHAUEN (ed.), "Methods and Applications in Adaptive Control," Springer-Verlag, Berlin and New York, 1980.

7. C. J. HARRIS and S. A. BILLINGS (eds.), "Self-Tuning and Adaptive Control: Theory and Applications," Peregrinus, London, 1981.

7a. G. C. GOODWIN and K. S. SIN, "Adaptive Filtering, Prediction and Control," Prentice-Hall, Englewood Cliffs, New Jersey, 1984.

8. H. T. TOIVONEN, *Model. Identif. Control 5*, 19 (1983).

9. V. PETERKA and K. J. ÅSTRÖM, *Proc. IFAC Symp. Identif. Syst. Parameter Estim., 3rd, The Hague* (1973).

10. U. BORISSON, "Self-Tuning Regulators—Industrial Applications and Multivariable Theory," Report, No. 7513, Department of Automatic Control, Lund Institute of Technology, Lund, Sweden, 1975.

11. U. BORISSON, *Automatica 15*, 209 (1979).

12. H. N. KOIVO, *Automatica 16*, 351 (1980).

13. L. KEVICZKY and K. S. P. KUMAR, *Int. J. Control 33*, 913 (1981).

14. K. J. ÅSTRÖM, "Introduction to Stochastic Control Theory," Academic Press, New York, 1970.

15. P. M. MÄKILÄ, T. WESTERLUND, and H. T. TOIVONEN, *Automatica 20*, 15 (1984).

16. M. ATHANS, *IEEE Trans. Autom. Control AC-16*, 529 (1971).

17. J. F. MACGREGOR, *Can. J. Chem Eng. 51*, 468 (1973).

18. T. SÖDERSTRÖM, L. LJUNG, and I. GUSTAVSSON, *IEEE Trans. Autom. Control AC-21*, 837 (1976).

19. P. E. CAINES and S. LAFORTUNE, *IEEE Trans. Autom. Control AC-29*, 312 (1984).

20. P. M. MÄKILÄ, *Optim. Control Appl. Methods 3*, 337 (1982).

21. L. LJUNG and T. SÖDERSTRÖM, "Theory and Practice of Recursive Identification," MIT Press, Cambridge, Massachusetts, 1983.

22. L. LJUNG, *Int. J. Control 27*, 673 (1978).

23. G. BIERMAN, "Factorization Methods for Discrete Estimation," Academic Press, New York, 1977.

24. V. PETERKA, *Kybernetika 11*, 53 (1975).

25. K. S. SIN and G. C. GOODWIN, *Automatica 18*, 315 (1982).

26. T. R. FORTESCUE, L. S. KERSHENBAUM, and B. E. YDSTIE, *Automatica 17*, 831 (1981).

27. T. HÄGGLUND, *Prepr. IFAC Workshop Adapt. Syst. Control Signal Process., San Francisco* (1983).

28. G. C. GOODWIN, P. J. RAMADGE, and P. E. CAINES, *SIAM J. Control Optim. 19*, 829 (1981).

29. L. LJUNG, *IEEE Trans. Autom. Control AC-22*, 551 (1977).

30. L. LJUNG, *Proc. Jt. Autom. Control Conf., 1980, San Francisco* (1980).

31. K. J. ÅSTRÖM and B. WITTENMARK, *Proc. IFAC Symp. Stochastic Control, Budapest* (1974).

32. L. LJUNG and B. WITTENMARK, "Asymptotic Properties of Self-Tuning Regulators," Report, No. 7404, Department of Automatic Control, Lund Institute of Technology, Lund, Sweden, 1974.

33. H. T. TOIVONEN, Ph.D. Thesis, Department of Chemical Engineering, Åbo Akademi, Åbo (Turku), Finland, 1981.

34. H. T. TOIVONEN, *Automatica 19*, 415 (1983).

35. K. P. LAM, "Implicit and Explicit Self-Tuning Controllers," OUEL Report, No. 1334/80, University of Oxford, Oxford, 1980.

36. K. J. ÅSTRÖM and Z. ZHAO-YING, *Proc. Workshop Adapt. Control, Florence, Italy* (1982).

37. H. EL-SHERIEF and N. K. SINHA, *Automatica 18*, 101 (1982).

38. L. S. SHIEH, C. T. WANG, and Y. T. TSAY, *Proc. IEE 130-D*, 143 (1983).

39. C. SAMSON and J. J. FUCHS, *Proc. IEE 128-D*, 102 (1981).

40. D. W. CLARKE, *Automatica 20*, 501 (1984).

41. D. W. CLARKE and P. J. GAWTHROP, *Automatica 17*, 233 (1981).

42. H. T. TOIVONEN, *Int. J. Control 38*, 229 (1983).

43. T. WESTERLUND, *IEEE Trans. Autom. Control AC-26*, 885 (1981).

44. H. T. TOIVONEN, submitted *Optimal Control Appl. Meth. 7*, 305 (1986).

45. H. PANOSSIAN and C. T. LEONDES, *Int. J. Syst. Sci. 14*, 385 (1983).

46. K. J. ÅSTRÖM and B. WITTENMARK, *Automatica 9*, 185 (1973).

47. G. C. GOODWIN, B. C. MCINNIS, and J. C. WANG, *Proc. 21st IEEE Conf. Decision Control, Orlando* (1982).

48. L. DUGARD, G. C. Goodwin, and X. XIANYA, *Automatica 20*, 701 (1984).

49. D. W. CLARKE and P. J. GAWTHROP, *Proc. IEE 122-D*,929 (1975).

50. D. W. CLARKE and P. J. GAWTHROP, *Proc. IEE 126-D*, 633 (1979).

51. P. E. MODÉN and T. SÖDERSTRÖM, *IEEE Trans. Autom. Control AC-27*, 214 (1982).

52. G. FAVIER and M. HASSANI, *Proc. 21st IEEE Conf. Decision Control, Orlando* (1982).

53. H. T. TOIVONEN, "A Self-Tuning Controller for Systems with Non-Square Transfer Functions," Report, No. 83-3, Process Control Laboratory, Åbo Akademi, Åbo (Turku), Finland, (1983).

54. G. MENGA and E. MOSCA, *in* "Advances in Control," p. 334 (D. G. Lainiotis and N. S. Tzannes, eds.), Reidel, Dordrecht, The Netherlands, 1980.

55. C. GRECO, G. MENGA, E. MOSCA, and G. ZAPPA, *Automatica 20*, 681 (1984).

A Covariance Control Theory

ANTHONY F. HOTZ
ROBERT E. SKELTON

School of Aeronautics and Astronautics
Purdue University
West Lafayette, Indiana 47906

I. INTRODUCTION

Covariance *analysis* permeates almost all of systems theory. In fact, covariance analysis has been the key tool in the development and maturity of several different problems in systems theory. Three important areas include identification, state estimation, and model reduction.

First, covariance analysis has been used in identification theory even prior to the modern renewal of state-space ideas [1-3]. When taking data from unknown systems, it is eminently practical to compute the first two moments of the stochastic process; the mean and the covariance. To compute fewer than two reduces to a study of mean values only. To compute more than two requires either knowledge of the distribution of the random vectors or much numerical computation. For many years, ARMA models and their covariances have enjoyed a wide acceptance as a practical means of identification [1-3]. Identification theory allows some specified structure (ARMA structure is one example), whereas the available realization theory makes no presumption of internal structure. Both theories rely on covariances that make up the Toeplitz matrix [4-7]. This theory is far from finalized, especially the search for a theory of minimal realizations for a given covariance sequence [6-7].

Second, the theory of state estimation and Kalman filtering promises certain properties of the covariance matrix of the estimation error [8]. Indeed, the entire development of optimal state estimation is based upon covariance analysis of stochastic processes.

Third, in recent years, model reduction theory has recognized the importance of covariances in establishing model error criteria. Researchers in model reduction have tried matching poles [9-10], Markov parameters [11-12], frequency response [13], and impulse response [14], but recently covariances have played a major role. Reduced order models that match the entire (infinite) covariance sequence are called stochastically equivalent realizations (SERs), and these realizations can be lower in order than Kalman's minimal realization (minimal observable, controllable dimension) [15]. Reduced order models that provide "closest" approximations to the infinite covariance sequence have been defined [16], as well as models that match a *specified* number of covariances [17-18].

It suffices to say, then, that there is ample motivation and theory for the *analysis* of covariances. But a theory for the *control* of covariances is strangely absent. This paper proposes that such a theory would be important to control engineering and offers some initial theorems.

The motivation for a theory to control covariances stems from two needs.

The first need (and the primary motivation of this study) is that many engineering systems have performance requirements that are naturally stated in terms of the root-mean-square (RMS) values, $\bar{y}_i \triangleq \left[\underline{E}\left(y_i^2\right)\right]^{1/2}$, of *each* of the n_y outputs of the system

$$\bar{y}_i \leq \sigma; \quad \sigma, \bar{y}_i \in \underline{R}^{n_y}. \tag{1}$$

For example, a spacecraft with a large antenna might need to *simultaneously* achieve a 1-mm RMS deflection at 20 points over the surface of the parabolic dish $\left(\bar{y}_i \leq 1 \text{ mm}, i = 1, \ldots, 20\right)$, pointed within 0.01° RMS to an earth receiver station $\left(\bar{y}_i \leq 0.01°, i = 21, 22\right)$, and an RMS value of strain of 0.001 at a critical point in the structure $\left(\bar{y}_{23} \leq 0.001\right)$. In this hypothetical example, the space mission may not be feasible unless the control system can achieve (1) with

$$\sigma^T = (1, 1, \ldots, 1, 0.01, 0.01, 0.001) \tag{2}$$

subject to a specified disturbance environment and control limitations (which might also be given in RMS terms, i.e., $\bar{u}_i < \mu_i$, $i = 1, n_u$). Unfortunately, there is presently no control theory to directly accommodate such straightforward design goals, and yet a large percentage of real-world objectives are of this nature. Two common approaches to problems of this type are linear quadratic (LQ) control theory and pareto-optimal control theory. Linear quadratic theory attempts to achieve (1) by minimizing a weighted *sum* of the performance objectives

$$\underline{V} = \sum_{i=1}^{n_y} q_i \bar{y}_i^2 + \sum_{i=1}^{n_u} r_i \bar{u}_i^2.$$

However, minimizing the sum does not guarantee that individual objectives will be met. Attempts have been made to find weights r_i and q_i in an iterative fashion to achieve the individual objectives, $\bar{y}_i \le \sigma_i$, $i = 1, \ldots, n_y$, and $\bar{u}_j \le \mu_j$, $j = 1, \ldots, n_u$, for specified values of σ_i and μ_j, but there is no guarantee that such weights exist. Even when they do exist, the available iterative schemes [19,20] have no proof of convergence.

Pareto-optimal control theory [21,22] treats the vector performance index (1) by considering the class of controls closest to the set of controls that minimize *each* performance index $\bar{y}_i^2$, subject to specified control limitations on each input, $\bar{u}_i$. The difficulties with this technique are its computationally intensive nature and the nonuniqueness of the controls.

A theory of covariance control would provide a direct approach for achieving performance goals of this type. For example, (2) could be achieved by assigning the diagonal elements of the output covariance matrix with values given by (2). Of course, one still has to assign the off-diagonal elements, but this freedom allows specifications for other feedback properties.

The second need for a theory of covariance control is to promote integration of modeling and control problems. Since identification, state estimation, and model reduction use covariances as a measure of performance, a theory to *control* covariances would allow one to address the entire class of problems (modeling and control) of concern in systems via the *same* measure of performance. This could have drastic impact in the theory of adaptive control, where the current identification methods and control methods are often based upon different measures of performance. This difference of performance measures, in part, accounts for the performance degradation that commonly occurs when systems are integrated. For example, if one criterion is used for model reduction, another for sensor/actuator location, another for control law design, and another for identification, then we have no right to expect good performance in the *integration* of these tasks. Since the integration of the various design steps is merely the juxtaposition of different mathematical problems, a unified performance criteria seems natural. Currently, little has been done even in the integration of any *two* of these tasks. It is hoped that a theory of covariance control will allow the proper integration of more, if not all, of the tasks involved in the control analysis, design, and synthesis. Linear quadratic Gaussian (LQG) theory has allowed the integration of some of these tasks by the use of a similar quadratic cost function, but the disadvantage is that the quadratic cost function is artificial.

As a first step toward satisfying these needs, this article develops a complete theory for designing linear feedback controllers so that the closed-loop system achieves a specified state covariance. In Section III, we assume that the complete state of the system can be accurately measured at all times and is available for feedback. Though this case has limited practical application, it does provide the essential foundation for the more important case of state covariance assignment using state estimate feedback, which is treated in Section V.

The article is organized as follows. Section II introduces some fundamental concepts of covariance control theory and offers a precise statement of the objectives of the ensuing theory. Sections III and V define the necessary and sufficient conditions for existence of constant linear feedback controllers that assign a specified state covariance to the closed-loop system and identify the resulting set of feedback gain matrices. Section III considers full-state feedback controllers, and Section V considers observer-based systems. It will be shown that when existence is verified, the set of solutions is, in general, very large; Section IV offers a strategy for choosing specific solutions from that set. Section VI draws a relationship between controllers that achieve state covariance assignment and LQR designs. Finally, Section VII closes with an example.

II. FUNDAMENTAL CONCEPTS AND PROBLEM STATEMENT

This article focuses on statistically stationary systems, that is, systems that are described by constant coefficient linear equations forced by stationary random processes. These systems enjoy wide application in many engineering disciplines and accurately represent many important physical systems. Good examples include the response of aircraft to wind disturbances, motion of a spacecraft disturbed by actuator noise and crew motion, navigation systems with noisy sensors, and flexible structures with noisy actuators.

However, though these systems are modeled as stationary processes, their true nature is nonlinear and time varying; so it would seem that a theory of covariance control for these systems should be developed first. On the other hand, linear control theory for time-invariant systems is highly mature and a theory of covariance control for these systems has immediate practical value. In either case, this theory provides a new design tool and requires some new definitions.

Problem Statement. For the stationary linear system described by

$$\text{(1)} \quad \dot{x}(t) = Ax(t) + Bu(t) + \Gamma w(t), \quad E(w(t)) = 0,$$

$$(2)\quad u(t) = Gx(t), \quad E(x_0 w(t)^T) = 0,$$

$$E(w(t)w(\tau)^T) = W\delta(t - \tau); \; W > 0,$$

$$|B^T B| \neq 0; \quad \Gamma = [B \; \Gamma'],$$

determine all state feedback gains G such that the state covariance

$$X \triangleq \lim_{t\to\infty} E(x(t)x(t)^T)$$

achieves a specified value $\overline{X}$.

Now x(t), u(t), and w(t) are n_x, n_u, n_w vectors, respectively, and A, B, Γ, G, W are $n_x \times n_x$, $n_x \times n_u$, $n_x \times n_w$, $n_u \times n_x$, and $n_w \times n_w$ matrices, respectively, with elements (constant) in the real field $\underline{R}$. We shall refer to this problem as state covariance assignment (SCA). We assume that the system (1), (2) is stabilizable and completely disturbable, that is, that (A, B) is stabilizable and (A, Γ) is a controllable pair. Also, as noted in the problem statement, we assume that the actuators are a disturbance source, that is, $\Gamma = [B \; \Gamma']$. This important assumption guarantees that (A + BG, Γ) is a controllable pair for any choice of G.

The relation defining the state covariance for (1), (2) is well established [23] and is given by the Liapunov equation,

$$(A + BG)X + X(A + BG)^T + \Gamma W \Gamma^T = 0. \tag{3}$$

Equation (3) allows the study of stationary covariances in a direct manner without requiring any integration of the associated power spectra. This equation has received much attention in recent years and its solution is routine [24,25]. However, the general Liapunov equation (3) is unknown in X, whereas in state covariance assignment (3) is unknown in G. That is, we assign a value to each entry of X and seek the feedback gain G that satisfies (3). Clearly, routine solution methods do not apply. In fact, the character of the equation is quite different. For example, it is well known that when (A + BG) is a stability matrix, (3) has a unique solution for all $\Gamma W \Gamma^T$ [23]. Moreover, if (A, Γ) is a controllable pair, then so is (A + BG, Γ), and $X = X^T$ is positive definite. Now consider (3) unknown in G. Let $\overline{X} = \overline{X}^T > 0$ and (A, Γ) be a controllable pair. The following lemma states that, in general, (A + BG) is not unique. The exception is the scalar case.

Lemma 1. Consider the matrix equation

$$\underline{A}X + X\underline{A}^T + Q = 0 \tag{4}$$

unknown in $\underline{A}$. Let $Q = Q^T \geq 0$ and $X = X^T > 0$. Then the solution $\underline{A}$ is, in general, not unique. The exception is the scalar case.

Proof. Barnett and Storey [26] recognized that (4) has the solution

$$\underline{A}X = -\frac{1}{2}Q + S_K, \tag{5}$$

where S_K is a skew symmetric matrix. When solving (5) for X, S_K is unique. However, when solving for $\underline{A}$, S_K is arbitrary since $\underline{A}$ does not demand symmetry. Hence $\underline{A} = \left(-\frac{1}{2}Q + S_K\right)X^{-1}$. For the scalar case, $S_K = 0$ and $\underline{A}$ is unique.

Lemma 1, though straightforward, is key to all subsequent theory developed in this chapter. It is instrumental in defining the set of all gain matrices that satisfy (3).

It is important to note that all solutions $\underline{A}$ to (4) are stability matrices. This fact trivially follows from the Liapunov stability theory and is formalized next.

Lemma 2 (Liapunov [27]). A matrix $\underline{A}$ is a stability matrix, that is, all eigenvalues of $\underline{A}$ have negative real parts, if and only if for any given $Q = Q^T > 0$ there exists an $X = X^T > 0$ that satisfies

$$\underline{A}X + X\underline{A}^T + Q = 0.$$

Lemma 3 (Kalman [27]). If $(\underline{A}, \sqrt{Q})$ is a controllable pair, then Lemma 2 holds with $Q = Q^T \geq 0$.

Using the previous three lemmas, we can define the set of all closed loop systems $\underline{A}$ that achieve a specified state covariance $\overline{X}$.

Theorem 1

For the stabilizable, completely disturbable, stationary system,

$$\dot{x} = \underline{A}x(t) + \Gamma w(t), \quad E(w(t)) = 0,$$

$$E(w(t)w(\tau)^T) = W\delta(t - \tau); \quad W > 0,$$

the set of all matrices $\underline{A}$ that achieve a specified state covariance, $\overline{X} = \lim_{t\to\infty} E(x(t)x(t)^T)$, $\overline{X} > 0$, is given by

$$\underline{\underline{A}} = \left\{\left(-\frac{1}{2}\,\Gamma W \Gamma^T + S_K\right)\overline{X}^{-1};\ S_K + S_K^T = 0\right\}. \tag{6}$$

Moreover, any $\underline{A} \in \underline{\underline{A}}$ is a stability matrix.

Proof. The proof follows immediately from Lemmas 1, 2, and 3.

It is clear that Theorem 1 falls short of achieving our goal. Having defined the set of closed-loop systems achieving a specified covariance $\overline{X}$ does not guarantee the existence of a suitable gain matrix G. In fact, it is quite possible that no solution exists. In other words, it is possible that the specified state covariance $\overline{X}$ cannot be achieved by any feedback gain G. This notion leads naturally to the important concept of covariance controllability.

Definition 1. The completely disturbable, stationary system (1) is said to be *completely state covariance controllable* if and only if there exists a control law (2) that transfers any initial state covariance $X_0 > 0$ in the limit as $t \to \infty$.

This definition naturally aligns with the notion of complete state controllability and the connection between these two concepts requires close examination. One might suspect that complete covariance controllability is a stronger system property than is complete state controllability. Indeed, this is the case and will be formalized shortly.

We now state the conditions for when a system (1) is completely state covariance controllable. Theorem 2 defines necessary and sufficient conditions and Corollaries 1, 2, and 3 interpret these conditions.

Theorem 2

The completely disturbable, stationary system

$$\begin{aligned} \dot{x} = Ax + Bu + \Gamma w, \quad & E(w(t)) = 0; \\ & E(w(t)w(\tau)^T) = W\,(t - \tau); \quad W > 0 \end{aligned} \tag{7}$$

is completely state covariance controllable if and only if

(i) $BB^+A = A$

and

(ii) $BB^+\Gamma W\Gamma^T = \Gamma W \Gamma^T$,

where B^+ denotes the Moore-Penrose inverse of B[1].

Proof. Suppose that (7) is completely state covariance controllable. Then there exists a gain matrix G such that

$$(A + BG)X + X(A + BG)^T + \Gamma W \Gamma^T = 0 \tag{8}$$

for any specified state covariance $\bar{X} = \bar{X}^T > 0$. Using Lemma 1,

[1] *See Section IX.*

(8) has the solution

$$(A + BG)\overline{X} = -\frac{1}{2}\,\Gamma W \Gamma^T + S_K, \qquad S_K + S_K^T = 0, \tag{9}$$

or

$$BG\overline{X} = -\frac{1}{2}\,\Gamma W \Gamma^T + S_K - A\overline{X}. \tag{10}$$

Equation (10) is a matrix equation unknown in G. From the theory of generalized inverses, it is well known that (10) has a solution if and only if (see Appendix A)

$$BB^+\left[-\frac{1}{2}\,\Gamma W \Gamma^T + S_K - A\overline{X}\right] = \left[-\frac{1}{2}\,\Gamma W \Gamma^T + S_K - A\overline{X}\right] \tag{11}$$

or

$$(I - BB^+)A\overline{X} = (I - BB^+)\left(-\frac{1}{2}\,\Gamma W \Gamma^T + S_K\right). \tag{12}$$

Equation (12) is independent of $\overline{X}$ if and only if

$$(I - BB^+)A = 0 \tag{13}$$

and

$$(I - BB^+)\left(-\frac{1}{2}\,\Gamma W \Gamma^T + S_K\right) = 0. \tag{14}$$

By hypothesis, (10) is consistent; hence (13) and (14) must be satisfied; that is, $BB^+A = A$ and $BB^+\left(-\frac{1}{2}\,\Gamma W \Gamma^T + S_K\right) = \left(-\frac{1}{2}\,\Gamma W \Gamma^T + S_K\right)$. It is not hard to show that S_K plays no role here. Simply use the fact that BB^+ is idempotent and that $\Gamma W \Gamma^T \geq 0$. The details are omitted for brevity. Hence the condition $BB^+(\Gamma W \Gamma^T) = \Gamma W \Gamma^T$.

Now suppose that $BB^+A = A$ and $BB^+\Gamma W \Gamma^T = \Gamma W \Gamma^T$. Then (12) is satisfied for any choice of $\overline{X} = \overline{X}^T > 0$ and any choice of S_K such that $BB^+S_K = S_K$ (note $S_K = 0$ is such a choice); hence (10) is consistent, that is, has a solution.

Remark. If $\Gamma = B$, then condition (ii) of Theorem 2 is always satisfied, leaving only the condition $BB^+A = A$.

Corollary 1 (a necessary condition). If the system (7) is completely state covariance controllable, then

$\text{rank}[A] \leq \text{rank}[B]$.

Proof. The rank of $\ker (I - BB^+) = \text{rank}(B)$. Hence $A \subset \ker(I - BB^+)$ implies $\text{rank}(A) \leq \text{rank}(B)$.

Corollary 2 (a sufficient condition). If $\text{rank}[B] = n_x$, then the system (7) is completely state covariance controllable.

Proof. Suppose that $\text{rank}[B] = n_x$; then $BB^+ = I$ and conditions (i) and (ii) of Theorem 2 are satisfied.

Corollary 3. If $\text{rank}[A] = n_x$ (or $\text{rank}[\Gamma W \Gamma^T] = n_x$), then the system (7) is completely state covariance controllable if and only if $\text{rank}[B] = n_x$.

Proof. Assume that $\text{rank}[A] = n_x$ and the system (7) is completely state covariance controllable. Then by Theorem 2, $BB^+A = A$, which by Corollary 1 implies $\text{rank}[B] = n_x$. Now suppose that $\text{rank}[B] = n_x$; then use Corollary 2. The same argument applies if $\text{rank}[\Gamma W \Gamma^T] = n_x$.

Corollary 3 has an interesting interpretation for mechanical systems. A mechanical system that has no zero frequency modes is completely state covariance controllable if and only if it has as many independent actuators as states.

The conditions for complete state covariance controllability defined in Theorem 2 emphasize the importance of actuator location. The condition $BB^+A = A$ stresses compatibility between actuator location and open-loop system dynamics and the condition $BB^+\Gamma W \Gamma^T = \Gamma W \Gamma^T$ identifies the actuator locations necessary to damp disturbances. Clearly, actuator location plays an important role in covariance controllability just as it does in state controllability. However, though these two controllability concepts have similar interpretations, they are not synonomous. Theorem 3 shows, as suspected earlier, that state covariance controllability is a stronger system property than is state controllability.

Theorem 3

If the system (1) is completely state covariance controllable, then (1) is completely state controllable. The converse is not necessarily true.

Proof. We prove the contrapositive. Suppose that (1) is not completely state controllable. Then, in controllable canonical form, the system (1) has the form

$$\begin{vmatrix} \dot{x}_{\bar{c}} \\ \dot{x}_{\bar{c}} \end{vmatrix} = \begin{vmatrix} \bar{A}_{cc} & \bar{A}_{c\bar{c}} \\ 0 & \bar{A}_{\overline{cc}} \end{vmatrix} \begin{vmatrix} x_c \\ x_{\bar{c}} \end{vmatrix} + \begin{vmatrix} \bar{B}_c \\ 0 \end{vmatrix} u(t) + \begin{vmatrix} \bar{\Gamma}_c \\ \bar{\Gamma}_{\bar{c}} \end{vmatrix} w(t).$$

Then $\bar{B}^+ = \left[\bar{B}_c^+ \;\; 0\right]$ and

$$\overline{BB}^+ A = \begin{vmatrix} \bar{B}_c\bar{B}_c^+ & 0 \\ 0 & 0 \end{vmatrix} \begin{vmatrix} \bar{A}_{cc} & \bar{A}_{c\bar{c}} \\ 0 & \bar{A}_{\overline{cc}} \end{vmatrix} \neq \begin{vmatrix} \bar{A}_{cc} & \bar{A}_{\overline{cc}} \\ 0 & \bar{A}_{\overline{cc}} \end{vmatrix},$$

which violates condition (i) of Theorem 2. Hence (1) is not state covariance controllable. We show that the converse is not true by counterexample. Let

$$A = \begin{vmatrix} 1 & 0 \\ 0 & 1 \end{vmatrix}, \quad B = \begin{vmatrix} 1 \\ 1 \end{vmatrix}.$$

Clearly, (A, B) is a controllable pair. However,

$$BB^+ = \begin{vmatrix} \frac{1}{2} & \frac{1}{2} \\ \frac{1}{2} & \frac{1}{2} \end{vmatrix};$$

hence $BB^+A \neq A$, which violates condition (i) of Theorem 2.

It is evident that the conditions for complete state covariance controllability are severe. In fact, most systems lack this property and the cost of achieving it is very high, for example, adding actuators. However, complete state covariance controllability is a much stronger property than what is really required. Performance objectives for engineering systems are specific and focus on a particular set of requirements. Hence, SCA is only concerned with achieving a *particular* state covariance. Clearly, the ability to achieve any state covariance is sufficient to meet this goal but certainly is not necessary. Apparently, state covariance uncontrollability does *not* preclude good covariance designs.

Since most systems are state covariance uncontrollable, it is more practical to define the necessary and sufficient conditions for the system (1) to achieve a particular state covariance. This is the topic of the next section.

III. MAIN THEOREMS ON STATE COVARIANCE ASSIGNMENT

In this section, we present the two main results of this chapter. Theorem 4 states the necessary and sufficient conditions for the existence of a linear state feedback control law (2) that will assign a particular state covariance $\overline{X}$ to the system (1). We assume that the complete state of the system can be accurately measured at all times and is available for feedback. Theorem 5 defines the complete set of solutions, that is, the set of gain matrices.

To provide for a clear exposition of the following theory, we make the coordinate transformation

$$x = T\hat{x},$$

where T is a factor of the specified covariance $\bar{X}$, that is, $\bar{X} = TT^T$. Though not particularly useful in this section, this coordinate change does enlighten us in the following sections and is made at this time for consistency. However, all results are stated in original coordinates. The theory that follows is not bound by this particular choice of coordinates nor by the particular choice of the factor T. Hence let T be *any* factor of $\bar{X}$ and define $\hat{A} = T^{-1}AT$, $\hat{B} = T^{-1}B$, $\hat{\Gamma} = T^{-1}\Gamma$, and $\hat{G} = GT$. Also, we assume that $\hat{B}$ is nonsquare but has linearly independent columns; that is, $\text{rank}(\hat{B}) = n_u$. This assumption merely states that we assume independent actuators. It too is nonessential, but provides for a clear presentation. Finally, because generalized inverses are used extensively in this section, a few basic facts and properties of these inverse are reviewed in Appendix A.

In transformed coordinates, the specified state covariance is the identity matrix $\hat{\bar{X}} = I$; hence (3) becomes

$$(\hat{A} + \hat{B}\hat{G}) + (\hat{A} + \hat{B}\hat{G})^T + \hat{\Gamma}W\hat{\Gamma}^T = 0. \tag{15}$$

Clearly, $\hat{\Gamma}W\hat{\Gamma}^T$ defines the symmetric part of the closed-loop system $(\hat{A} + \hat{B}\hat{G})$. Following the arguments of Lemma 1, (15) has the solution

$$\hat{A} + \hat{B}\hat{G} = -\frac{1}{2}\,\hat{\Gamma}W\hat{\Gamma}^T + S_K, \tag{16}$$

where S_K defines the skew symmetric part of $\hat{A} + \hat{B}\hat{G}$. Then (16) can be written as

$$\hat{B}\hat{G} = -\frac{1}{2}[\hat{\Gamma}W\hat{\Gamma}^T + (\hat{A} + \hat{A}^T)] + S_K, \tag{17}$$

where S_K now defines only the skew symmetric part of $\hat{B}\hat{G}$.

Consistency of equation (17) clearly depends, in part, upon choice of S_K. This freedom can be exploited to a limited degree in forcing consistency of (17). Let $\hat{B}^+$ be the Moore-Penrose inverse of $\hat{B}$. Then (17) is consistent if and only if

$$\hat{B}\hat{B}^+\left[-\frac{1}{2}(\hat{\Gamma}W\hat{\Gamma}^T + \hat{A} + \hat{A}^T) + S_K\right] = \left[-\frac{1}{2}(\hat{\Gamma}W\hat{\Gamma}^T + \hat{A} + \hat{A}^T) + S_K\right] \tag{18}$$

or

$$(I - \hat{B}\hat{B}^+)\tilde{S}_K = (I - \hat{B}\hat{B}^+)(\hat{A} + \hat{A}^T + \hat{\Gamma}W\hat{\Gamma}^T) \tag{19}$$

for some skew-symmetric matrix $\tilde{S}_K = 2S_K$. Equation (19) clearly shows the importance of properly choosing S_K. The reader should be cautioned against the erroneous conclusion that consistency of (17) can always be forced by proper choice of $\tilde{S}_K$. This is certainly not true. This is the subject of Theorem 4, which states the necessary and sufficient conditions for the existence of solutions to (17).

Theorem 4

For the stabilizable, completely disturbable, time-invariant linear system,

$$\dot{x} = Ax + Bu + \Gamma w, \qquad E(w(t)) = 0,$$
$$u = Gx, \qquad E(w(t)w(\tau)^T) = W\delta(t - \tau); \quad W > 0,$$

there exists a gain matrix G such that the closed-loop system achieves a specified state covariance $\overline{X}$ if and only if

$$[F^T(\hat{A} + \hat{A}^T + \hat{\Gamma}W\hat{\Gamma}^T)F]_{22} = [0], \tag{20}$$

where $[\cdot]_{22}$ is the 22 block of $[\cdot]$ and has dimension $(n_x - n_u) \times (n_x - n_u)$, F is the modal matrix of $\hat{B}\hat{B}^+$, and $\hat{A} = T^{-1}AT$, $\hat{B} = T^{-1}B$, $\hat{\Gamma} = T^{-1}\Gamma$, $\bar{X} = TT^T$.

Proof. Consistency of (17) depends on the existence of $\tilde{S}_K$ such that (19) is satisfied. We note that $\hat{B}\hat{B}^+$ and $(I - \hat{B}\hat{B}^+)$ are idempotent, symmetric matrices (see Appendix A). Let F be the unitary modal matrix of $\hat{B}\hat{B}^+$. Then premultiplying both sides of (19) by F^T and postmultiplying both sides by F yields

$$F^T(I - \hat{B}\hat{B}^+)FF^T\tilde{S}_K F = F^T(I - \hat{B}\hat{B}^+)FF^T(\hat{A} + \hat{A}^T + \Gamma W\Gamma^T)F. \tag{21}$$

Now,

$$FF^T = F^TF = I, \quad F^T\tilde{S}_K F = - F^T\tilde{S}_K F^{\;T} = \bar{S}_K,$$

and

$$F^T\hat{B}\hat{B}^+F = \left|\begin{matrix} I_{n_u} & 0 \\ 0 & 0 \end{matrix}\right|.$$

Then (21) becomes

$$\left|\begin{matrix} 0 & 0 \\ 0 & I_{(n_x - n_u)} \end{matrix}\right| \left|\begin{matrix} \bar{S}_{K_{11}} & \bar{S}_{K_{12}} \\ -\bar{S}^T_{K_{12}} & \bar{S}_{K_{22}} \end{matrix}\right| = \left|\begin{matrix} 0 & 0 \\ 0 & I_{(n_x - n_u)} \end{matrix}\right| \left|\begin{matrix} \underline{D}_{11} & \underline{D}_{12} \\ \underline{D}^T_{12} & \underline{D}_{22} \end{matrix}\right|, \tag{22}$$

where we have defined

$$F^T(\hat{A} + \hat{A}^T + \hat{\Gamma}W\hat{\Gamma}^T)F = \underline{D}.$$

With these preliminaries, we now prove the theorem. Suppose that there exists a gain G that achieves the state covariance $\bar{X}$. Then there exists a skew symmetric matrix $\tilde{S}_K$ that satisfies (19) and hence (22). Expanding (22) reveals that $\bar{S}_{K_{11}}$ is an arbitrary $n_u \times n_u$ skew-symmetric matrix, $\bar{S}_{K_{12}} = -\underline{D}_{12}$

and $\bar{S}_{K_{22}} = \underline{D}_{22}$. Now, $\bar{S}_{K_{22}}$ is skew symmetric while $\underline{D}_{22}$ is symmetric. Thus $\bar{S}_{K_{22}} = \underline{D}_{22}$ is possible if and only if $\underline{D}_{22} = [0]$; hence

$$\underline{D}_{22} = [F^T(\hat{A} + \hat{A}^T + \hat{\Gamma}W\hat{\Gamma}^T)F]_{22} = [0].$$

Now suppose that $[F^T(\hat{A} + \hat{A}^T + \hat{\Gamma}W\hat{\Gamma}^T)F]_{22} = [0]$. Then from (22), any choice of skew-symmetric $\bar{S}_{K_{11}}$ yields the skew-symmetric matrix

$$\tilde{S}_K = F\begin{vmatrix} \bar{S}_{K_{11}} & -\underline{D}_{12} \\ \underline{D}_{12}^T & 0 \end{vmatrix}F^T \tag{23}$$

that satisfies (19). Hence (17) is consistent and there exists a gain (or gains) G that achieve $\bar{X}$.

The following corollary states in the original coordinates the condition of Theorem 4. Though somewhat less insightful, it is easier to compute. However, the size of the resulting test matrix may be greatly increased.

Corollary. Condition (20) is equivalent to

$$(I - BB^+)(\Gamma W\Gamma^T + A\bar{X} + \bar{X}A^T)(I - BB^+) = 0.$$

Proof. Condition (20) can be written as

$$\begin{vmatrix} 0 & 0 \\ 0 & I_{(n_x-n_u)} \end{vmatrix} F^T(\hat{\Gamma}W\hat{\Gamma}^T + \hat{A} + \hat{A}^T)F \begin{vmatrix} 0 & 0 \\ 0 & I_{(n_x-n_u)} \end{vmatrix} = 0.$$

This is equivalent to

$$F^T(I - \hat{B}\hat{B}^+)FF^T(\hat{\Gamma}W\hat{\Gamma}^T + \hat{A} + \hat{A}^T)(I - \hat{B}\hat{B}^+)F = 0$$

or

$$(I - BB^+)(\hat{\Gamma}W\hat{\Gamma}^T + A + A^T)(I - BB^+) = 0,$$

which in original coordinates yields the desired result.

The final theorem of this section defines the set of all gain matrices such that the system (1) achieves the specified state covariance $\bar{X}$ when solutions do exist.

Theorem 5

Consider the stationary system given in Theorem 4. If the conditions of Theorem 4 are satisfied, then the set of all gain matrices G such that the closed-loop system achieves the assigned state covariance $\bar{X}$ is given by

$$\underline{G} = \left\{ G : G = -\frac{1}{2}\hat{B}^{+}\left(\hat{\Gamma}W\hat{\Gamma}^{T} + \hat{A} + \hat{A}^{T} - \tilde{S}_{K}\right)T^{-1} \right.$$
$$\left. = -\frac{1}{2}B^{+}\left(\Gamma W\Gamma^{T} + A\bar{X} + \bar{X}A^{T} - T\tilde{S}_{K}T^{T}\right)\bar{X}^{-1} \right\}, \qquad (24)$$

where $\tilde{S}_K$ satisfies (23) and

$$\bar{X} = TT^{T}, \quad \hat{A} = T^{-1}AT, \quad \hat{B} = T^{-1}B, \quad \hat{\Gamma} = T^{-1}\Gamma.$$

Moreover, all $G \in \underline{G}$ are stabilizing.

Proof. By hypothesis, Theorem 4 is satisfied; hence (24) is true for any $\tilde{S}_K$ of the form (23). Equation (24) is the general solution to equation (12) assuming $|\hat{B}\hat{B}^{T}| \neq 0$ and so defines *every* solution to (17). The fact that every $G \in \underline{G}$ is stabilizing follows immediately from Lemmas 1, 2, and 3.

Remark. It is of interest to note that $T\tilde{S}_K T^T$ can be written as

$$T\tilde{S}_{K}T^{T} = \bar{F}\begin{vmatrix} S_{K_{11}} & -D_{12} \\ D_{12}^{T} & 0 \end{vmatrix}\bar{F}^{T},$$

where $S_{K_{11}}$ is an arbitrary $n_u \times n_u$ skew-symmetric matrix, $\bar{F}$ is the modal matrix of BB^{+}, and $\bar{F}^{T}(\Gamma W\Gamma^{T} + A\bar{X} + \bar{X}A^{T})\bar{F} = D$. This removes from the set of solutions any requirement of the coordinate transformation T.

From Theorem 5, we see that for every choice of skew-symmetric $\bar{S}_{K_{11}}$, the matrix $\tilde{S}_k$ [as defined by (23)] generates an element of $\underline{G}$. For single-input systems, $\bar{S}_{K_{11}} = 0$ and the gain G that achieves $\bar{X}$ is unique. However, for multiple-input systems $\bar{S}_{K_{11}}$ is arbitrary (skew symmetric); hence for this case $\underline{G}$ is a very large set. This fact is not surprising since SCA is a multiobjective design task. Unlike scalar objective designs that often lead to unique solutions, such as linear quadratic techniques in which only one gain matrix minimizes a scalar cost functional, multiobjective designs often lead to nonunique solutions, for example, pareto-optimal control theory.

The nonuniqueness of the submatrix $\bar{S}_{K_{11}}$ allows some additional design freedom by providing a large set of possible gain matrices. This freedom can be exploited to achieve other desired closed-loop properties such as low control input requirements, good transient behavior, and reduced sensitivity to modeling error. Only the former of these tasks is explored in this article.

The focus of the next section is on choosing $\bar{S}_{K_{11}}$ to minimize the upper bound of the mean-square amplitude of each control input. This will allow a priori specification of actuator power requirements necessary to achieve the assigned state covariance.

IV. CONTROL ENERGY CONSIDERATIONS

The reader will recognize that assigning the state covariance of a system is only half of the problem. Ideally, one would like to assign the state covariance matrix and the input covariance matrix simultaneously. In that way, complete system performance requirements can be specified. However, this objective is very demanding and in many cases unnecessarily complicates the design task. Instead, since the set $\underline{G}$ contains every possible gain matrix that achieves the specified state covariance $\overline{X}$, it is more practical to find a $G_* \in \underline{G}$ that best satisfies the control input specifications.

Here, we want to find a $G_* \in \underline{G}$ that minimizes the mean-square amplitude of each actuator input in the following sense.

Define α_i as the actuator size requirement for the ith actuator. Also, define the input covariance

$$U \triangleq \lim_{t\to\infty} E(u(t)u(t)^T) = GXG^T$$

and the set of input covariances $\underline{U}$ as

$$\underline{U} = \left\{G\overline{X}G^T : G \in \underline{G}\right\}.$$

The ith diagonal entry of any $U \in \underline{U}$ represents the mean-square amplitude of the input to the ith actuator and the $G_* \in \underline{G}$ we seek generates a corresponding $U_* \in \underline{U}$ such that for each $u_{ii} \in \mathrm{diag}(U_*)$,

$$u_{ii} \leq \alpha_i. \tag{25}$$

We will find that, in general, G_* is not unique. The remainder of this section pursues two tasks: (1) to evaluate α_i for each actuator and (2) to find G_*.

Assume Theorem 4. Using (23) and (24) we can write

$$F^T\hat{B}\hat{G}F = -\frac{1}{2}\,\underline{D} + \frac{1}{2}\,F^T\tilde{S}_kF$$

$$= -\frac{1}{2}\begin{vmatrix} \underline{D}_{11} & \underline{D}_{12} \\ \underline{D}_{12}^T & 0 \end{vmatrix} + \frac{1}{2}\begin{vmatrix} \overline{S}_{K_{11}} & -\underline{D}_{12} \\ \underline{D}_{12}^T & 0 \end{vmatrix}, \tag{26}$$

where F, $\underline{D}$, and $\tilde{S}_k$ are as previously defined. Then, multiplying (26) by its transpose yields

$$F^T\hat{B}\hat{G}FF^T\hat{G}^T\hat{B}^TF = F^T\hat{B}\hat{G}\hat{G}^T\hat{B}^TF$$

$$= \begin{vmatrix} \frac{1}{4}\left(\underline{D}_{11} - \bar{S}_{K_{11}}\right)\left(\underline{D}_{11} - \bar{S}_{K_{11}}\right)^T + \underline{D}_{12}\underline{D}_{12}^T & 0 \\ 0 & 0 \end{vmatrix}. \quad (27)$$

Observe that $\hat{G}\hat{G}^T = U \in \underline{U}$, the input covariance, and that

$$F^T\hat{B} = \begin{vmatrix} \beta \\ 0 \end{vmatrix},$$

where β is $n_u \times n_u$. Equation (27) reveals that

$$\beta U \beta^T = \frac{1}{4}\left(\underline{D}_{11} - \bar{S}_{K_{11}}\right)\left(\underline{D}_{11} - \bar{S}_{K_{11}}\right)^T + \underline{D}_{12}\underline{D}_{12}^T. \quad (28)$$

This important equation relates the input covariance U to the skew-symmetric matrix $\bar{S}_{K_{11}}$, which is free to be chosen. We recall that $\bar{S}_{K_{11}}$ generates the set of gains $\underline{G}$, which in turn generates the set of input covariances $\underline{U}$. However, (28) relates $\bar{S}_{K_{11}}$ and U directly. Our present aim is to use (28) to evaluate the ASR for each actuator.

Rearranging (28) and equating only the diagonal elements yields the equation

$$\left[4U - 4\beta^{-1}\underline{D}_{12}\underline{D}_{12}^T\beta^{-1^T}\right]_{ii} = \left[\beta^{-1}\left(\underline{D}_{11} - \bar{S}_{K_{11}}\right)\right. \quad (29)$$

$$\left. \times \left(\underline{D}_{11} - \bar{S}_{K_{11}}\right)^T\beta^{-1^T}\right]_{ii}. \quad (29)$$

Let $\underline{C} = \beta^{-1}\underline{D}_{12}\underline{D}_{12}^T\beta^{-1^T}$ and $\underline{K} = \beta^{-1}\left(\underline{D}_{11} - \bar{S}_{K_{11}}\right)$. Define c_{ii} as the ith diagonal element of $\underline{C}$, k_i^T as the ith row of $\underline{K}$, and b_i as the ith row of β^{-1}. Then (29) can be written

$$4u_{ii} - 4c_{ii} = k_i^T k_i \quad (30)$$

or

$$u_{ii} = \frac{k_i^T k_i}{4} + c_{ii}. \tag{31}$$

Apparently, the mean-square amplitude of the input to the ith actuator depends only on the length (squared) of the row vector k_i^T and the constant (positive) c_{ii}. However, though c_{ii} is known and fixed, $k_i^T k_i$ is subject to choice of $\bar{S}_{K_{11}}$; hence we seek a relationship between the length of k_i and $\bar{S}_{K_{11}}$.

It is easy to see that k_i is given by

$$k_i = \left(\underline{D}_{11} + \bar{S}_{K_{11}}\right) b_i^T; \tag{32}$$

then

$$||k_i|| = ||\left(\underline{D}_{11} + \bar{S}_{K_{11}}\right) b_i^T||, \tag{33}$$

where $||\cdot||$ denotes the euclidian norm of $(\cdot)$.

Clearly, the length of k_i is but the magnification of the length of b_i^T by $\left(\underline{D}_{11} + \bar{S}_{k_{11}}\right)$. This is a useful interpretation for it relates naturally to the spectral norm of $\left(\underline{D}_{11} + \bar{S}_{K_{11}}\right)$, which measures the maximum magnification of the length of any vector x by the transformation $\left(\underline{D}_{11} + \bar{S}_{K_{11}}\right)$.

Recall that the spectral norm of a matrix is defined as its maximum singular value $\bar{\sigma}(\cdot)$. Then

$$\bar{\sigma}\left(\underline{D}_{11} + \bar{S}_{K_{11}}\right) = \max_{||x|| \neq 0} \frac{||\left(D_{11} + \bar{S}_{K_{11}}\right)x||}{||x||},$$

or we can write, using (33),

$$\bar{\sigma}\left(\underline{D}_{11} + \bar{S}_{K_{11}}\right) \geq \frac{||\left(\underline{D}_{11} + \bar{S}_{K_{11}}\right) b_i^T||}{||b_i^T||} \tag{34}$$

for all b_i^T. We note that since $|\beta| \neq 0$, $||b_i^T|| \neq 0$ for all i.

Finally, (33), (34) yield the inequality

$$\|k_i\| \leq \left(\bar{\sigma}\left(\underline{D}_{11} + \bar{S}_{K_{11}}\right)\right)\|b_i^T\|$$

or using (31)

$$u_{ii} \leq \left\{\frac{\left(\bar{\sigma}\left(\underline{D}_{11} + \bar{S}_{K_{11}}\right)\right)\|b_i^T\|}{2}\right\}^2 + c_{ii}. \tag{35}$$

Equation (35) defines an upper bound on the mean-square amplitude of each actuator input for any choice of $\bar{S}_{K_{11}}$. Therefore, we can now find the least upper bound for each u_{ii}; that is, we can determine α_i as

$$\alpha_i = \min_{\bar{S}_{K_{11}}}\left\{\left(\frac{\left[\bar{\sigma}\left(\underline{D}_{11} + \bar{S}_{K_{11}}\right)\right]\|b_i^T\|}{2}\right)^2 + c_{ii}\right\}. \tag{36}$$

As stated earlier, $(ASR)_i$ is a measure of the control authority required to achieve the specified state covariance. We now add the interpretation that if the ith actuator is *not capable* of responding to an input with a RMS value of $\sqrt{\alpha_i}$, then there is *no guarantee* that the actuator is "large" enough to achieve the assigned state covariance.

Evaluating (36) requires that we find $\bar{S}_{K_{11}}$ such that $\bar{\sigma}\left(\underline{D}_{11} + \bar{S}_{K_{11}}\right)$ is minimized. This is the subject of the following lemma.

Lemma 4. Let E be a given diagonal matrix and let $\underline{S}$ be the set of all skew symmetric matrices. Then

$$\min_{S \in \underline{S}} \bar{\sigma}(E + S) = \max_i |e_{ii}|,$$

where $e_{ii} \in \text{diag}(E)$.

Proof. Observe the structure of (E + S),

$$(E + S) = \begin{vmatrix} e_{11} & s_{12} & \cdots & s_{1n} \\ -s_{12} & e_{22} & \cdots & \cdot \\ -s_{1n} & \cdot & \cdots & e_{nn} \end{vmatrix}.$$

It is easy to see that $\Sigma_i |a_{ij}| = \Sigma_i |a_{ji}|$ for all j, where $a_{ij} \in (E + S)$. In words, the sum of the absolute value of the elements of corresponding rows and columns is equal. Hence the matrix norms $||E + S||_1 = \max_j \Sigma_i |a_{ij}|$ and $||E + S||_\infty = \max_i \Sigma_j |a_{ij}|$ are equal. This allows us to exploit the following matrix norm inequalities:

$$\bar{\sigma}^2(E + S) \leq ||E + S||_\infty ||E + S||_1. \tag{37}$$

$$||E + S||_A = \max_{i,j} |a_{ij}| \leq \bar{\sigma}(E + S). \tag{38}$$

Now $||E + S||_1 = ||E + S||_\infty$, and so (37) and (38) yield the

$$||E + S||_A \leq \bar{\sigma}(E + S) \leq ||E + S||_\infty \tag{39}$$

or

$$\max_{ij} |a_{ij}| \leq \bar{\sigma}(E + S) \leq \max_i \sum_j |a_{ij}|. \tag{40}$$

Observe that $\Sigma_i |a_{ij}| = |e_{ii}| + \Sigma_i |s_{ij}|$. Then if we let $|e_{11}| = \max_i |e_{ii}|$ and choose $|s_{i1}| = |s_{1i}| = 0$ for all i and choose all other $|s_{ij}|$ such that

$$|e_{jj}| + \sum_j |s_{ij}| \leq |e_{11}| \quad \text{for all} \quad j \quad (j \neq 1),$$

then it follows that

$$\min_{S \in \underline{S}} ||E + S||_\infty = |e_{11}|$$

or

$$\min_{S \in \underline{S}} \bar{\sigma}(E + S) \leq |e_{11}|.$$

On the other hand, it is easy to see that

$$\min_{S \in \underline{S}} \|E + S\|_A = |e_{11}|;$$

hence (40) yields the desired result.

We cannot use Lemma 4 directly since, in general, $\underline{D}_{11}$ is not diagonal. However, the singular values of a matrix are invariant under unitary transformations, and so we let M be the unitary modal matrix of $\underline{D}_{11}$ and define $E = M^T \underline{D}_{11} M$ and $S = M^T \bar{S}_{K_{11}} M$. Then applying Lemma 4, (36) becomes

$$\alpha_i = \left\{\frac{(\max_i |e_{ii}|) \|b_i^T\|}{2}\right\}^2 + c_{ii}, \tag{41}$$

where $e_{ii} \in \operatorname{diag}\left(M^T \underline{D}_{11} M\right)$.

Equation (41) achieves the first task of this section. The final task is to find $G_* \in \underline{G}$. This is done in Theorem 6.

Theorem 6

Let $\underline{G}$ be the set of all gain matrices that achieve an assigned state covariance $\bar{X}$. Let $\underline{U}$ be the corresponding set of input covariance matrices. Let $\underline{D}_{11}$ be as defined in Theorem 4. Then all akew-symmetric matrices of the form

$$\tilde{S}_K = F \begin{vmatrix} MSM^T & -\underline{D}_{12} \\ \underline{D}_{12}^T & 0 \end{vmatrix} F^T$$

generate the set of gain matrices $\underline{G}_* \in G$,

$$\underline{G}_* = \left\{G_* : G_* = -\frac{1}{2}\left(\hat{\Gamma} W \hat{\Gamma} + \hat{A} + \hat{A}^T - \tilde{S}_K\right) T^{-1}\right\}$$

such that each $G_* \in \underline{G}$ has the property: For every $U_* \in \underline{U}_* \subset \underline{U}$,

$$\underline{U}_* = \left\{G_* \bar{X} G_*^T;\ G_* \in \underline{G}_*\right\},$$

each $u_{ii} \in \text{diag}(U_*)$ satisfies the condition

$$u_{ii} \leq \alpha_i,$$

where S is skew symmetric with the following properties: If $|e_{11}| = \max_i |e_{ii}|$, $e_{ii} \in \text{diag}\left(M^T \underline{D}_{11} M\right)$, then $|s_{j1}| = |s_{1j}| = 0$ for all j and $|e_{jj}| + \Sigma_j |s_{ij}| \leq |e_{11}|$ for all i, j ($\neq 1$). Here M is the modal matrix of $\underline{D}_{11}$, F is the modal matrix of $\hat{B}\hat{B}^T$, $\overline{X} = TT^T$, and b_i^T, c_{ii} are as previously defined.

Proof. The proof follows immediately from Lemma 4, (35), and (41).

It is interesting to note that S = [0] generates a gain matrix in the set $\underline{G}_*$. If no other design objectives are required, this is a simple and convenient solution. It is also of interest to note that $\underline{G}_*$ is a large set and therefore allows some extra design freedom without compromising either the state covariance assignment or the boundedness property of the diagonal elements of U_*.

Finding the set $\underline{G}_*$ was rather circuitous; however, the result is very important since it emphasizes a practical strategy for choosing a particular gain from the set $\underline{G}$. We close this section with a theorem that revisits the idea of simultaneously assigning both the state covariance and the input covariance.

Theorem 7

Let $\underline{G}$ be the set of all gain matrices such that the closed-loop system achieves an assigned state covariance $\overline{X}$. Let $\underline{U}$ be the set of all corresponding input covariance matrices. Let $\overline{U}$ be an assigned input covariance matrix. Then $\overline{U} \in \underline{U}$ if and only if there exists a matrix Z such that $4\left(\beta\overline{U}\beta^T - \underline{D}_{12}\underline{D}_{12}^T\right) = ZZ^T$ and $\frac{1}{2}(Z + Z^T) = \underline{D}_{11}$, where β, $\underline{D}_{11}$, $\underline{D}_{12}$ are as previously defined.

Proof. Suppose that $\overline{U} \in \underline{U}$. Then using (28),

$$4\left(\beta\overline{U}\beta^T - \underline{D}_{12}\underline{D}_{12}^T\right) = \left(\underline{D}_{11} - \overline{S}_{K_{11}}\right)\left(\underline{D}_{11} - \overline{S}_{K_{11}}\right)^T.$$

Let $Z = \underline{D}_{11} - \overline{S}_{K_{11}}$; then

$$\frac{1}{2}(Z + Z^T) = \frac{1}{2}\left\{\left(\underline{D}_{11} - \overline{S}_{K_{11}}\right) + \left(\underline{D}_{11} + \overline{S}_{K_{11}}\right)\right\} = \underline{D}_{11}.$$

Now suppose that

$$4\left(\beta\overline{U}\beta^T - \underline{D}_{12}\underline{D}_{12}^T\right) = ZZ^T$$

and

$$\frac{1}{2}(Z + Z^T) = \underline{D}_{11}$$

for some matrix Z. Then define the skew symmetric matrix $-S = Z - \underline{D}_{11}$; hence

$$4\left(\beta\overline{U}\beta^T - \underline{D}_{12}\underline{D}_{12}^T\right) = (\underline{D}_{11} - S)(\underline{D}_{11} - S)^T,$$

which has the form (28); S then generates the required $G \in \underline{G}$ such that $G\overline{X}G^T = \overline{U}$.

Remark. The fact that the factor of $4\left(\beta\overline{U}\beta^T - \underline{D}_{12}\underline{D}_{12}^T\right)$ is not unique makes Theorem 6 computationally difficult. This topic is still under investigation.

V. STATE COVARIANCE ASSIGNMENT USING STATE ESTIMATE FEEDBACK

In this section, we formulate the state covariance assignment problem under the assumption that the complete state vector cannot be measured and the measurements that are available are corrupted by sensor noise. Even so, we will see that under certain conditions the results from the previous sections are well tailored to solve the problem here as well.

We consider the stable, time-invariant, linear system described by

$$\dot{x}(t) = Ax(t) + Bu(t) + \Gamma w(t) \qquad E(w(t)) = E(v(t)) = 0$$

$$z(t) = Mx(t) + v(t) \qquad E(x_0 w(t)) = E\left(x_0 v(t)^T\right) = 0$$

$$u(t) = G\hat{x}(t) \qquad E(w(t)w(\tau)^T) = W(t-\tau);\ W > 0 \tag{42}$$

where (A, B), (A, Γ) and (A, C) are respectively stabilizable, controllable, and detectable. Let the estimator assume the standard form,

$$\dot{\hat{x}}(t) = A\hat{x}(t) + Bu(t) + K(z(t) - M\hat{x}(t))$$

where $\hat{x}$ is an n_x vector of estimates of the states, z is an n_p vector of measurements of the outputs of the system, K is an $n_x \times n_p$ matrix of estimator gains, M is an $n_p \times n_x$ measurement matrix and A, B, G, and u are as previously defined.

An obvious area of concern is how to best choose K and G to achieve the control objectives. One might suggest that a Kalman filter be employed to obtain an "optimal" estimate of the states. On the other hand, one might gain advantage by implementing a suboptimal filter. In this way a tradeoff between noise rejection and certain desirable covariance properties may be realized. Advantages of adding "fictitious" noise sources (which result in suboptimal filters) to compensate for model uncertainties have been shown [30,31]. Such studies emphasize the need to carefully match estimator performance with the control objectives. This topic is beyond the scope of this paper. The use of dynamic controllers for state covariance assignment is under investigation.

In this paper we let K be the Kalman gain, $K = PM^TV^{-1}$, where P is the state estimation error covariance and satisfies,

$$(A - KM)P + P(A - KM)^T + PM^TV^{-1}MP + \Gamma W\Gamma^T = 0.$$

Indeed, the smallest error variance for each state is given by the Kalman filter.

The associated Liapunov equation becomes [25]

$$(A + BG)(\overline{X} - P) + (\overline{X} - P)(A + BG)^T + KVK^T = 0 \qquad (43)$$

where $(\overline{X} - P) > 0$.

Conveniently, (43) is similar to (8) and it would appear that Theorems 4 and 5 of Section III are applicable to (43) as well. However, it is not clear that all choices of G satisfying (43) will yield a stable system, i.e., there is no guarantee that (A + BG, K) will be a controllable pair for all choices of G. A similar situation was encountered in Section II for the state feedback case, where the problem was remedied by assuming noisy actuators. A similar mathematical device is employed here.

Let $\overline{K} = [K|B]$ and

$$\overline{V} = \begin{vmatrix} V & 0 \\ 0 & V' \end{vmatrix}, \qquad V' = \epsilon I$$

where $\epsilon > 0$ is an arbitrarily small number. This amounts to adding a fictitious forcing function (with arbitrarily small norm) to (43) and guarantees that $(A + BG, \overline{K})$ is a controllable pair for all G. Then, by virtue of the added ϵ term (43) becomes

$$(A + BG)(\overline{X} - P) + (\overline{X} - P)(A + BG)^T + KVK^T + \epsilon BB^T = 0 \quad (44)$$

and hence all G satisfying (44) are stabilizing. Consequently, using the G computed via (44) in (43) yields

$$(A + BG)(\overline{X} - \Delta X - P) + (\overline{X} - \Delta X - P)(A + BG)^T + KVK^T = 0 \tag{45}$$

where ΔX appears by removing the fictitious forcing function.

Subtracting (44) from (45) results in

$$(A + BG)\Delta X + \Delta X(A + BG)^T + \epsilon BB^T = 0. \tag{46}$$

Now, $(A + BG)$ is stable and $(A + BG, B)$ is a controllable pair, hence from Liapunov stability theory, $\Delta X > 0$ and consequently $\overline{X} - \Delta X < \overline{X}$.

This tells us that though $\overline{X}$ is no longer assigned exactly, the rms values, $[\overline{X} - \Delta X]_{ii}$, *still* satisfy the design constraints, $E\left(x_i^2\right) \leq [\overline{X}]_{ii}$, for all $i = 1, 2, \ldots, n_x$.

Conveniently, Theorems 4 and 5 of Section III are directly applicable to (44). Note that we just consider the system (42) and replace $\Gamma W \Gamma^T$ by $\overline{KVK^T}$ and $\overline{X}$ by $(\overline{X} - P)$ appropriately. Also, all results of Section IV carry over as well.

VI. CONNECTIONS TO LINEAR QUADRATIC CONTROL THEORY

The linear quadratic regulator problem (LQR) in optimal control theory needs little introduction. This design technique is perhaps the most important development in modern systems theory, as evidenced by the enormity of literature on the subject and also by its appearance as a major focus in most systems theory texts written after 1960 [23,27,30-32].

The beauty of LQR design lies in its simplicity, intuitive appeal, and stability properties. It gives one a systematic procedure for computing constant feedback control gains for multiple-input systems by way of a scalar performance criteria that, in general, is a function of control effort and system output response.

However, one should recognize that this scalar cost functional is a device rather than a measure of performance. The underlying objective of regulator design is to keep a randomly disturbed, stationary system within an acceptable deviation from a reference state using acceptable control activity. Hence an appropriate measure of system performance is the RMS amplitudes of the system states or outputs. In fact, LQR designs often require several iterations (weighting adjustments) before acceptable RMS responses are achieved. In a sense, then, one might say that LQR optimal control theory is an indirect approach to state covariance assignment. The purpose of this section is to explore the relationship between these two design techniques.

Specifically, we want to determine whether LQR design methodology imposes constraints on covariance design. That is, are there viable covariance matrices that cannot be achieved by LQR techniques?

One approach to this problem is to determine which control gains in the set $\underline{G}$ (the set of all gains achieving an assigned state covariance) are optimal with respect to a quadratic cost functional. This question is certainly not new. Kalman [33] first investigated this "inverse problem of optimal control" in 1964. There he considered only single-input systems. In 1966, Anderson [34] extended the results to multiple-input systems. Many papers followed [35-37] as interest in this problem peaked in the early 1970s.

We begin by noting that Theorem 5 defines the set of all gain matrices that achieve an assigned state covariance. Hence any LQR design that achieves the same state covariance must be in that set. Consider the completely controllable system (1) and the quadratic cost functional

$$J = \frac{1}{2}\int_0^\infty (x^T Q x + u^T R u)\, dt, \tag{47}$$

where Q, R are symmetric "weighting" matrices with constant coefficients. The feedback control law G^0 that minimizes (47) subject to (1) is well known. It is given by

$$G^0 = -R^{-1}B^T P, \tag{48}$$

where P is the solution of the steady-state matrix Riccati equation

$$PA + A^T P - PBR^{-1}B^T P + Q = 0. \tag{49}$$

It is also well understood that $R > 0$ is a necessary condition for the existence of a unique minimizing control u and that $Q \geq 0$ is sufficient to guarantee existence or a solution P to (49). Here we assume only that $R > 0$. Although $Q \geq 0$ is a commonly adopted policy, it is unnecessarily restrictive for the problem posed here.

It should be noted that if (47) is generalized to include a cross-product term $2u^T N x$, then every stabilizing $G \in \underline{G}$ is optimal [40]. However, in practice, use of this term is uncommon and is somewhat difficult to physically motivate. Here we restrict our discussion to the case $N = 0$. We shall see that this restriction is not unduly severe.

We now present three lemmas followed by the main theorem of this section.

Lemma 5 [37]. Given any stabilizing gain matrix G, (48) has solutions $R = R^T > 0$ and $P = P^T$ if and only if GB has linearly independent real eigenvectors and rank BG = rank G.

Proof. See [37].

Lemma 6 [37]. Consider the stable closed-loop linear system (1), (2). It is possible to construct a performance index (47) with $Q = Q^T$, $R = R^T > 0$ that attains its absolute minimum I^0 over all square integrable controls for all x_0 if and only if GB has linearly independent real eigenvectors and rank BG = rank G.

Proof. See [37].

Remark. Note that since B is assumed to have linearly independent columns, then the rank condition rank BG = rank G is always satisfied.

Lemma 7. The matrix products $BR^{-1}B^T$ and BB^+ share the same modal matrix, where $R = R^T > 0$ and B^+ is the Moore-Penrose inverse of B.

Proof. Let R be any positive definite symmetric matrix. Let T be the unitary modal matrix of $BR^{-1}B^T$. Then

$$T^TBR^{-1}B^TT = \begin{vmatrix} D_R & 0 \\ 0 & 0 \end{vmatrix}, \tag{50}$$

where D_R is $n_u \times n_u$ and diagonal with real entries. Now, $BB^+BR^{-1}B^T = BR^{-1}B^T$; hence $T^TBB^+TT^TBR^{-1}B^TT = T^TBR^{-1}B^TT$. Then using (50)

$$T^TBB^+T\begin{vmatrix} D_R & 0 \\ 0 & 0 \end{vmatrix} = \begin{vmatrix} D_R & 0 \\ 0 & 0 \end{vmatrix}. \tag{51}$$

Let

$$T^T BB^+ T = \begin{vmatrix} Z_{11} & Z_{12} \\ Z_{12}^T & Z_{22} \end{vmatrix}.$$

Then (51) yields the equations

$$Z_{11} D_R = D_R, \; Z_{12}^T D_R = 0$$

from which it follows that $Z_{11} = I$, $Z_{12}^T = 0$. So

$$T^T BB^+ T = \begin{vmatrix} I_{n_u} & 0 \\ 0 & Z_{22} \end{vmatrix}$$

and since $\text{rank}(BB^+) = \text{rank}(B)$, then $\text{rank}(Z_{22}) = 0$ or $Z_{22} = 0$. So,

$$T^T BB^+ T = \begin{vmatrix} I_{n_u} & 0 \\ 0 & 0 \end{vmatrix}$$

and T is the modal matrix of BB^+.

Note that Lemmas 5 and 6 simply state that for a given gain matrix G if there exists an $R > 0$ and $P = P^T$ satisfying (48), then G minimizes a quadratic cost functional of the form (47). There are many such results in the literature, with varying degrees of generality. Lemmas 5 and 6 are well suited for the problem here. The following theorem shows that if the set of gain matrices $\underline{G}$ is not empty, then there exists (nonuniquely) a gain matrix G that simultaneously achieves a specified state covariance and minimizes a quadratic cost functional.

Theorem 8

Let $\underline{G}$ be the set of all gain matrices that achieve a specified state covariance for the system (1). Assume $\underline{G} \neq \phi$. Then for some $Q = Q^T$, $R = R^T > 0$, there exists a $G \in \underline{G}$ that minimizes a quadratic cost functional

$$J = \frac{1}{2}\int_0^\infty (x^TQx + u^TRu)\ dt.$$

Moreover, for any $R > 0$, $R = R^T$ and some $Q = Q^T$ there exists a corresponding $G \in \underline{G}$.

Proof. The unique minimizing control law of J subject to (1) is given by $G^0 = -R^{-1}B^TP$, where $R = R^T > 0$ and P satisfies (49). Again we work from the coordinate system where $\overline{X} = I$. Assume that $\hat{B}$ has linearly independent column vectors. If we can show that there exists a symmetric matrix $\overline{P}$ such that

$$-\hat{B}R^{-1}\hat{B}^T\overline{P} - \overline{P}\hat{B}R^{-1}\hat{B}^T + (\hat{\Gamma}W\hat{\Gamma}^T + \hat{A} + \hat{A}^T) = 0, \tag{52}$$

then it will follow from Theorem 5 and Lemmas 5 and 6 that $G^0 \in \underline{G}$. Choose any $R = R^T > 0$ and let F be the modal matrix of $BR^{-1}B^T$. Then we can write

$$-F^T\hat{B}R^{-1}\hat{B}^TFF^T\overline{P}F - F^T\overline{P}FF^T\hat{B}R^{-1}\hat{B}^TF + F^T(\hat{\Gamma}W\hat{\Gamma}^T + \hat{A} + \hat{A}^T)F = 0$$

or

$$\begin{vmatrix} D_R & 0 \\ 0 & 0 \end{vmatrix}\begin{vmatrix} \tilde{P}_{11} & \tilde{P}_{12} \\ \tilde{P}_{12}^T & \tilde{P}_{22} \end{vmatrix} + \begin{vmatrix} \tilde{P}_{11} & \tilde{P}_{12} \\ \tilde{P}_{12}^T & \tilde{P}_{22} \end{vmatrix}\begin{vmatrix} D_R & 0 \\ 0 & 0 \end{vmatrix} = \begin{vmatrix} \underline{D}_{11} & \underline{D}_{12} \\ D_{12}^T & \underline{D}_{22} \end{vmatrix}. \tag{53}$$

Equation (53) yields the equations

$$D_R\tilde{P}_{11} + \tilde{P}_{11}D_R = \underline{D}_{11} \tag{54a}$$

$$D_R\tilde{P}_{12} = \underline{D}_{12}, \tag{54b}$$

$$\underline{D}_{22} = 0. \tag{54c}$$

Since $\underline{G} \neq \phi$, Lemma 7 and Theorem 4 guarantee $\underline{D}_{22} = 0$. Since D_R is diagonal and nonsingular, it is easy to see that $\tilde{P}_{11}$ is symmetric and unique and $\tilde{P}_{12} = D_R^{-1}\underline{D}_{12}$. Note that $D_R > 0$. Then,

$$F^T\hat{B}R^{-1}\hat{B}^TFF^T\overline{P}F = -F^T\hat{B}\hat{G}F = \begin{vmatrix} D_R\tilde{P}_{11} & D_R\tilde{P}_{12} \\ 0 & 0 \end{vmatrix} \tag{55}$$

is unique; hence $\hat{G}$ is unique [since $\ker(\hat{B}) = 0$]. Now only $\tilde{P}_{22}$ is unspecified. It is immediate from (55) that

$$\hat{G}F = -(F^T\hat{B})^+ \begin{vmatrix} D_R\tilde{P}_{11} & D_R\tilde{P}_{12} \\ 0 & 0 \end{vmatrix} = -R^{-1}\hat{B}^T FF^T \bar{P}F$$

or

$$[\beta^+ \quad 0] \begin{vmatrix} D_R\tilde{P}_{11} & D_R\tilde{P}_{12} \\ 0 & 0 \end{vmatrix} = R^{-1}[\beta^T \quad 0] \begin{vmatrix} \tilde{P}_{11} & \tilde{P}_{12} \\ \tilde{P}_{12}^T & \tilde{P}_{22} \end{vmatrix}$$

from which it is clear that $\hat{G}F$ is invariant to choice of $\tilde{P}_{22}$. So choose any $\tilde{P}_{22} = \tilde{P}_{22}^T$. Then compute $P = T^{-1^T}\bar{P}T^{-1} = T^{-1^T}F\tilde{P}F^T T^{-1}$ and Q via

$$Q = -PA - A^TP + PBR^{-1}B^TP.$$

The triple (R, Q, P) satisfies Lemmas 5 and 6 and $G \in \underline{G}$.

Observe that no sign definiteness has been established on either P or Q. In fact, we can show that P can always be chosen nonpositive definite (and nonpositive semidefinite). On the same note, we can show that under certain conditions P can always be chosen positive definite. This freedom results from the unconstrained nature of the submatrix $\tilde{P}_{22}$ as noted in the preceding proof. This fact in itself is of little consequence since stability is not at issue here. However, it does lead to some interesting conclusions. For example, for a given covariance specification, if for any $R > 0$ there does not exist a $P > 0$ corresponding to any $G \in \underline{G}$, then this indicates that the commonly adopted restriction $Q \geq 0$ [A, $Q^{1/2}$) an observable pair] precludes achievement of the given covariance matrix by LQR design. Additionally, even for those designs with corresponding $P > 0$, there is no guarantee that there exists an associated $Q \geq 0$.

On the other hand, if for some diagonal $R > 0$ there exists a $Q > 0$ corresponding to some $G \in \underline{G}$, then G not only achieves a specified covariance design, but also exhibits the impressive robustness properties as LQR designs [38,39].

Since the issues of the previous discussion center around the existence of $P > 0$, it is of interest to define the necessary and sufficient conditions for when there exists a $P > 0$ corresponding to some $G \in \underline{G}$. As noted in the proof to Theorem 8, $\tilde{P}_{22}$ can be chosen arbitrarily subject only to symmetry. However, $\tilde{P}_{11}$ and $\tilde{P}_{12}$ are constrained by equations (54a) and (54b) and so dictate the sign definiteness of $\tilde{P}$, hence P. In fact, it is easy to see that only P_{11} determines the sign definiteness of P. Here we use the conditions [42] that $\tilde{P} > 0$ iff (i) $\tilde{P}_{11} > 0$ and (ii) $\tilde{P}_{22} - \tilde{P}_{12}^T\tilde{P}_{11}^{-1}\tilde{P}_{12} > 0$. Clearly, for any $\tilde{P}_{12}$, $\tilde{P}_{11}$ we can choose $\tilde{P}_{22}$ satisfying (ii), leaving then only the condition that $\tilde{P}_{11} > 0$. Now, $\tilde{P}_{11}$ is the unique solution to equation (54a). Note that $D_R > 0$ and diagonal and is uniquely determined by choice of R. Also, as $\underline{D}_{11}$ is a submatrix of $F^T(\hat{\Gamma}W\hat{\Gamma}^T + \hat{A} + \hat{A}^T)F$, no sign definiteness is guaranteed on $\underline{D}_{11}$. It would appear, then, that for any choice of $R > 0$, little could be said a priori of the sign definiteness of $\tilde{P}_{11}$. Of course, if $\underline{D}_{11} < 0$, then $\tilde{P}_{11} > 0$ is guaranteed (since $D_R > 0$). This follows directly from the Liapunov stability theory. However, the following lemma shows that the sign of $\tilde{P}_{11}$ (hence $\tilde{P}$) can be manipulated at will by the proper choice of R under mild conditions on $\underline{D}_{11}$.

Lemma 8. Consider the matrix equation

$$AX + XA^T = D, \tag{56}$$

where $A > 0$ and diagonal. Then for any $D = D^T$ there exists an

A such that $X > 0$ iff

$$\text{diag}[D] > 0.$$

Proof. See Appendix B.

We can apply Lemma 8 directly to (54a). The required $R > 0$ achieve $\tilde{P}_{11} > 0$ (provided $\text{diag}[\underline{D}_{11}] > 0$) is given by

$$R = \left\{\beta^{+} D_R \beta^{+T}\right\}^{-1},$$

where $D_R = \text{diag}\{\lambda_1, \ldots, \lambda_{n_n}\}$, λ_i is as provided by the construction offered in the proof to Lemma 8, and β is as previously defined. We now summarize the main point of this discussion.

Theorem 9

Given a specified state covariance matrix $\overline{X}$ and the system (1), (2) achieving that covariance, if any diagonal element of the submatrix

$$[F^T(\hat{\Gamma}W\hat{\Gamma}^T + \hat{A} + \hat{A}^T)F]_{11}$$

is less than or equal to zero, then there does not exist an $R > 0$ and $Q \geq 0$ such that

$$G^0 = -R^{-1}B^T P \in \underline{G},$$

where P is the positive definite solution to (5) and $\underline{G}$, F, $\hat{\Gamma}$, and $\hat{A}$ are as previously defined.

Proof. Immediate by Lemma 8 and the preceding discussion.

Evidently, certain covariance designs can never be achieved via standard [$R > 0$, $Q \geq 0$, $(A, Q^{1/2})$ an observable pair] LQR techniques. However, no claim is made as to whether such designs are desirable. This is a topic of future research.

Also, one should note that the matrix $F^T(\hat{\Gamma}W\hat{\Gamma}^T + \hat{A} + \hat{A}^T)F$ contains a wealth of important information: for example, see Theorems 4, 6, and 9.

Finally, Theorems 8 and 9 invite a natural question. Are there any gains $G \in \underline{G}$ that are not LQR solutions for any $Q = Q^T$. The answer is yes, and we show this as follows. Lemmas 5 and 6 imply the existence of symmetric P as necessary. In proving Theorem 8 we found that $\tilde{P}_{22}$ can be chosen arbitrarily; therefore it can always be chosen symmetric, leaving only $\tilde{P}_{11}$ in question. Using (54a) we can write the solution of $\tilde{P}_{11}$ as

$$D_R P_{11} = -\frac{1}{2} \underline{D}_{11} + \bar{S}_{K_{11}}, \tag{57}$$

where $\bar{S}_{K_{11}}$ is the unique skew-symmetric matrix such that $\tilde{P}_{11}$ is symmetric. Note that D_R depends only on choice of R (and B) and that different choices of R will generally require a different skew-symmetric matrix $\bar{S}_{K_{11}}$ such that $\tilde{P}_{11}$ remains symmetric. Now recall that in the general theory of SCA, the set $\underline{G}$ is generated by choices of the matrix $\bar{S}_{K_{11}}$; hence we pose the following inverse problem: For any choice of skew-symmetric $\bar{S}_{K_{11}}$, does there exist a diagonal $D_R > 0$ and a symmetric $\tilde{P}_{11}$ such that (57) is satisfied? If true, then every $G \in \underline{G}$ is an LQR solution. (Note that this question is quite different from that posed in Theorem 9.) So, choose any $\bar{S}_{K_{11}}$ (skew symmetric). Using (57) we can write

$$\tilde{P}_{11} = D_R^{-1}\left(-\frac{1}{2} \underline{D}_{11} + \bar{S}_{K_{11}}\right), \tag{58}$$

and forcing symmetry on $\tilde{P}_{11}$ we arrive at the condition

$$D_R^{-1}\left(-\frac{1}{2} \underline{D}_{11} + \bar{S}_{K_{11}}\right) = \left(-\frac{1}{2} \underline{D}_{11} - \bar{S}_{K_{11}}\right)D_R^{-1}. \tag{59}$$

Equation (59) is both a necessary and sufficient condition for determining whether a given $G \in \underline{G}$ is also an LQR solution.

Theorem 10

Let $\underline{G}$ be the set of all gain matrices that achieve an assigned state covariance. Assume $\underline{G} \neq \phi$. Then any $G \in \underline{G}$ is optimal with respect to the cost function (47) for some $R = R^T > 0$ and $Q = Q^T$ if and only if there exists a diagonal matrix $D_R > 0$ such that

$$D_R^{-1}\left(-\frac{1}{2}\,\underline{D}_{11} + \bar{S}_{K_{11}}\right) = \left(-\frac{1}{2}\,\underline{D}_{11} - \bar{S}_{K_{11}}\right)\underline{D}_R^{-1}, \tag{60}$$

where $\bar{S}_{K_{11}}$ is the particular skew-symmetric matrix that generates G and $\underline{D}_{11}$ is as previously defined.

Proof. Assume that G is optimal from some $R = R^T > 0$ and $Q = Q^T$. Then using Lemma 5, P must be symmetric and hence (60) must be satisfied. Now suppose that there exists the required $D_R > 0$ satisfying (60). Then there exists the corresponding $R = R^T > 0$ and $P = P^T$ satisfying Lemmas 5 and 6.

It is not difficult to show by counterexample that the required D_R of Theorem 10 does not always exist. Therefore, we conclude that not every $G \in \underline{G}$ is LQR optimal. As a final note, recall from the previous section that for the case $\bar{S}_{K_{11}} = 0$ each actuator response is minimized in the sense defined there. For this case, (59) reduces to

$$D_R^{-1}\left[-\frac{1}{2}\,\underline{D}_{11}\right] = \left[-\frac{1}{2}\,\underline{D}_{11}\right]D_R^{-1} \tag{61}$$

for which $D_R = I$ is a corresponding solution; hence this gain

$$G|_{\bar{S}_{K_{11}}=0} \in \underline{G}_*$$

is an LQR solution. However, it is important to note that many gains in the set $\underline{G}_*$ cannot be generated via LQR techniques. In conclusion, we remark that every state covariance matrix generated via LQR design can be generated via SCA but not conversely.

The next section presents an example outlining an SCA design procedure.

VII. DESIGN EXAMPLE

In this section, a simple design example is presented that illustrates a few of the essential ideas of SCA.

Suppose that we wish to design a roll attitude regulator for a missile disturbed by random roll torques [39]. The control objective is to keep roll attitude small while staying within the physical limits of aileron deflection and aileron deflection rate.

We estimate the RMS value of the disturbing torque to be equivalent to 5° of aileron deflection and the correlation time to be 0.23 sec. The system is modeled as

$$\begin{vmatrix} \dot{\delta} \\ \dot{\omega} \\ \dot{\phi} \end{vmatrix} = \begin{vmatrix} 0 & 0 & 0 \\ 10 & -1 & 0 \\ 0 & 1 & 0 \end{vmatrix} \begin{vmatrix} \delta \\ \omega \\ \phi \end{vmatrix} + \begin{vmatrix} 1 \\ 0 \\ 0 \end{vmatrix} u(t) + \begin{vmatrix} 0 \\ 1 \\ 0 \end{vmatrix} v(t),$$

where δ is aileron deflection, ω is roll angular velocity, ϕ is roll angle, u is command signal to the aileron actuators, and v(t) is a white noise source with intensity $Q = 2((10)(5))^2(0.23$ $= 1150\ \text{deg}^2/\text{sec}^3$.

By specifying the state covariance, we can assign the desired RMS values for δ, ω, and ϕ. However, though we cannot directly assign an RMS value to $\dot{\delta}$, it is implicit in the specifications of the other variables. For example, if we desire a very small RMS value for roll angle, then the missile dynamics must be very fast, resulting in large RMS roll velocities and aileron deflection rates. Also, since this system is not disturbable in the actuator state, we must take care when assigning covariance values. With these considerations, we assign the RMS values for δ, ω, and ϕ as

$$\sqrt{E(\delta^2)} = 11 \text{ deg}, \quad \sqrt{E(\phi^2)} = 1 \text{ deg}, \quad \sqrt{E(\omega^2)} = 9 \text{ deg/sec}.$$

Assigning the desired cross correlations requires some insight. Since the angle ϕ cannot react instantaneously to changes in the angular velocity ω, there is no statistical dependence between ω *as a random variable* and ϕ at any given time in the steady state. Hence we assign $E(\phi\omega) = 0$. Assigning $E(\delta\phi)$ and $E(\omega\delta)$ requires a little more work, but not unreasonably so. We note that

$$\dot{\omega} = 10\delta - \omega + v(t). \tag{62}$$

Then multiplying (62) by ω and taking the expected value yields

$$E(\dot{\omega}\omega) = 10E(\delta\omega) - E(\omega^2) + E(\omega v)$$

or

$$E(\delta\omega) = \frac{1}{10}\{E(\dot{\omega}\omega) + E(\omega^2) - E(\omega v)\}.$$

Now $E(\dot{\omega}\omega) = 0$ by previous arguments and by definition $E(\omega v) = Q/2 = 575$. Then $E(\delta\omega) = -49.4$. Assigning $E(\delta\phi)$ is most easily achieved by using the consistency condition (12). Here we find that $E(\delta\phi) = -\frac{1}{10} E(\omega^2) = -8.1$.

The assigned state covariance is then

$$\overline{X} = \begin{vmatrix} 121.0 & -49.4 & -8.1 \\ -49.4 & 81.0 & 0 \\ -8.1 & 0 & 1.0 \end{vmatrix}.$$

Substituting A, B, Γ, and $\overline{X}$ into the consistency equation (12) yields the necessary skew-symmetric matrix

$$S_K = \begin{vmatrix} 0 & -1259.4 & 49.4 \\ 1259.4 & 0 & -81.0 \\ -49.4 & 81.0 & 0 \end{vmatrix}$$

and (10) gives the desired gain

$$G = [-14.56 \quad -24.43 \quad -68.57].$$

The input covariance is

$$E(u^2) = E(\dot{\delta}^2) = G\bar{X}G^T = 27379.8 \quad (\text{deg/sec})^2$$

or

$$\sqrt{E(u^2)} = 165.47 \quad \text{deg/sec}.$$

Indirect design approaches, such as LQR, generally require many adjustments of the weighting matrices before acceptable RMS values are achieved. Here, no iterations were necessary since the desired RMS values were specified.

Note that since this system has only one input, G is unique. However, for multiple-input systems G will not be unique.

The purpose of this example was to provide motivation for assigning the state covariance, particularly the cross-coupling terms that often require careful consideration. No particular algorithm has been offered, nor has the full power of the theory been exploited. For example, the consistency condition (12) was used as a guide to determine one of the cross-correlation terms. In general, however, one might require a particular state covariance matrix from which Theorem 4 will indicate consistency and Theorem 5 will define the set of solutions.

VIII. CONCLUSIONS

This article has introduced a theory for designing feedback controllers that assigns a specified state covariance to the closed loop system. The theory is restricted to linear time-invariant systems with constant gain state feedback or state-estimate feedback controllers.

The primary contributions of this article are summarized in Definition 1 and Theorems 2, 4, and 5.

Definition 1 and Theorem 2 define the notion of state covariance controllability. Theorem 4 characterizes all state covariance matrix values that a linear stochastic process may possess. Theorem 5 identifies the set of all constant state feedback gain matrices that achieve the assigned covariance.

When solutions exist, Theorem 5 provides a large set of possible gain matrices (except for single input systems where the solution is unique). Theorem 6 defines a strategy for choosing a particular gain matrix from that set so that an upper bound on the mean squared values of *each* control is minimized.

Furthermore, it was shown that when a Kalman filter is used to estimate the states, that the Theorems in both Sections III and IV apply with only minor modification.

It is well known that minimizing *scalar* quadratic functions leads to Riccati equations. So, it is fortunate and quite useful that a control which assigns *multiple* performance criteria (the entire state covariance matrix) can be obtained from a set of linear equations and therefore is even more tractable than the nonlinear (Riccati) equations.

Finally, it was shown that when solutions exist, that there exists gain matrices that simultaneously achieve a specified state covariance and minimize a quadratic cost functional (Theorem 8). Theorems 8, 9, and 10 conclude that every state covariance matrix generated via LQR design can be generated via SCA but not conversely.

APPENDIX A

This appendix reviews some basic facts on Moore-Penrose inverses and presents a well-established lemma on the solution of the matrix equation $AXB = D$. See [43] for an extensive treatment.

For every finite matrix A, there exists a unique matrix satisfying the following "Penrose" equations:

(1) $AXA = A$,

(2) $XAX = X$,

(3) $(AX)^* = AX$,

(4) $(XA)^* = XA$,

where the asterisk denotes conjugate transposition. The matrix X is known as the Moore-Penrose inverse A^+. Some useful properties of A^+ are listed below:

(i) If A is nonsingular, $A^+ = A^{-1}$ uniquely.

(ii) rank $A^+ \geq$ rank A.

(iii) AA^+ and A^+A are idempotent and have the same rank as A.

(iv) Let A be $m \times n$; then $A^+A = I_n$ iff rank $A = n$.

(v) Let A be $m \times n$; then $AA^+ = I_m$ if rank $A = m$.

(vi) $(A^+)^* = (A^*)^+$.

(vii) $(A^+)^+ = A$.

Lemma. Let $A \in C^{m\times n}$, $B \in C^{p\times q}$, $D \in C^{m\times q}$. Then the matrix equation $AXB = D$ is consistent if and only if for any† $A^{(1)}$, $B^{(1)}$

$$AA^{(1)}DB^{(1)}B = D,$$

†*The original theorem uses "some" in place of "any." In fact, if this condition holds for some $A^{(1)}$, $B^{(1)}$, then it holds for any $A^{(1)}$, $B^{(1)}$, as the reader can easily check.*

in which case the general solution is

$$X = A^{(1)}DB^{(1)} + Y - A^{(1)}AYBB^{(1)}$$

for arbitrary $Y \in C^{n\times p}$.

Then $A^{(1)}$ is a generalized inverse satisfying only the first "Penrose" equation. Clearly, we may replace $A^{(1)}$, $B^{(1)}$ by A^+, B^+ in the above lemma since these are also (1) inverse.

APPENDIX B: PROOF TO LEMMA 8

Given some $D = D^T$, suppose that there exists an $A > 0$ and diagonal and an $X > 0$ satisfying (1). Then diag[D] = 2 diag[AX]. Clearly diag[AX] > 0, and hence diag[D] > 0. Now let $A = \text{diag}\{\lambda_1 \cdots \lambda_n\}$ and

$$D = \begin{vmatrix} d_{11} & \cdots & d_{1n} \\ \cdot & \cdots & \cdot \\ \cdot & \cdots & \cdot \\ \cdot & \cdots & \cdot \\ \cdot & \cdots & \cdot \\ d_{1n} & \cdots & d_{nn} \end{vmatrix}$$

with $d_{ii} > 0$, $i = 1, \ldots, n$. Then X is given by

$$X = \begin{vmatrix} \dfrac{d_{11}}{2\lambda_1} & \dfrac{d_{12}}{\lambda_1 + \lambda_2} & \cdots & \dfrac{d_{1n}}{\lambda_1 + \lambda_n} \\ \cdot & \cdot & \cdots & \cdot \\ \cdot & \cdot & \cdots & \cdot \\ \cdot & \cdot & \cdots & \cdot \\ \dfrac{d_{1n}}{\lambda_1 + \lambda_n} & \cdot & \cdots & \dfrac{d_{nn}}{2\lambda_n} \end{vmatrix}. \tag{B1}$$

We want to show that for arbitrary $d_{ij} \in R$, there exists λ_i such that $X > 0$, provided that $d_{ii} > 0$ for all $i = 1, \ldots, n$. The proof is by induction, and we use the following properties of positive definite matrices:

(i) $X > 0$ iff the principle minors are all positive.

(ii) $X > 0$ iff $x^T X x > 0$ for all nonzero $x \in R^n$, that is, $\Sigma_{i=1}^n |x_i| < \infty$. Now for $k = 1$, the first principle minor, clearly $d_{11} > 0$ is all that is required. Let $k = 2$, the second principl minor; expanding the determinant reveals the condition

$$\frac{2\lambda_1\lambda_2}{(\lambda_1 + \lambda_2)^2} < \frac{d_{11}d_{22}}{d_{12}^2},$$

which is easily satisfied by the proper choice of λ_1, λ_2. Now suppose that the mth principle minor is positive. We want to show that we can shoose λ_{m+1} such that the $m + 1$ principle minor is positive. Proceeding by way of determinants is unwieldy. Instead, we show that $x^T X_{m+1} x > 0$ for all $x \in R^{m+1}$, where X_{m+1} is the $m + 1$ principle minor of X. Let $x^T = [v^T \; w]$, where $v \in R$ and $w \in R$. Since $X_m > 0$ by hypothesis, then $v^T X_m v > 0$ for all $v \in R^m$. Observe that

$$x^T X_{m+1} x = v^T X_m v + 2wv^T d + w^2 d_{m+1,m+1}/2\lambda_{m+1}, \tag{B2}$$

where d is the first m elements of the $m + 1$ column of X. We desire λ_{m+1} such that

$$v^T X_m v + 2mv^T d + w^2 d_{m+1,m+1}/2\lambda_{m+1} > 0$$

for all $v \in R^m$, $w \in R$. Clearly, only $2wv^T d$ may be negative; hence it is sufficient that

$$v^T X_m v + \frac{w^2 d_{m+1,m+1}}{2\lambda_{m+1}} > |2wv^T d| \tag{B3}$$

Using the Cauchy-Schwartz inequality, if we choose λ_{m+1} such that

$$v^T X_m v + \frac{w^2 d_{m+1,m+1}}{2\lambda_{m+1}} > 2|w| \ ||v|| \ ||d||; \tag{B4}$$

then (B3) is satisfied.

Using the fact that

$$\sigma_{min}(X_m)\,||v||^2 \leq v^T X_m v \leq \sigma_{max}(X_m)\,||v||^2,$$

where $\sigma_{max}(X_m)$, $\sigma_{min}(X_m)$ are the maximum and minimum eigenvalues of X_m, respectively, we write (B4) as

$$\sigma_{max}(X_m)\,||v||^2 + \frac{w^2 d_{m+1,m+1}}{2\lambda_{m+1}} > 2|w| \ ||v|| \ ||d||. \tag{B5}$$

Note that

$$d^T = \left[\frac{d_{1,m+1}}{\lambda_1 + \lambda_{m+1}}, \ldots, \frac{d_{m,m+1}}{\lambda_m + \lambda_{m+1}}\right].$$

Let $\lambda_{m+1} \gg \lambda_i$ for all i. Then

$$d^T \approx \left[\frac{d_{1,m+1}}{\lambda_{m+1}} \cdots \frac{d_{m,m+1}}{\lambda_{m+1}}\right] = \frac{1}{\lambda_{m+1}}[d_{i,m+1}, \ldots, d_{m,m+1}]$$

and

$$||d|| = \frac{1}{\lambda_{m+1}}||\bar{d}||.$$

Hence (B5) becomes

$$\sigma_{max}(X_m)\lambda_{m+1}||v||^2 + \frac{w^2 d_{m+1,m+1}}{2} - 2|w| \ ||v|| \ ||\bar{d}|| > 0 \tag{B6}$$

or

$$[\,||v|| \ \ |w|\,]\begin{bmatrix} \lambda_{m+1}\sigma_{max}(X_m) & \\ -||\bar{d}|| & \frac{d_{m+1,m+1}}{2} \end{bmatrix}\begin{bmatrix} ||v|| \\ |w| \end{bmatrix} > 0.$$

Then, if we choose

$$\lambda_{m+1} > \frac{2||\bar{d}||^2}{\sigma_{max}(X_m) d_{m+1,m+1}},$$

$\lambda_{m+1} \gg \lambda_i$ for all i, then $X_{m+1} > 0$ and the proof is complete.

REFERENCES

1. K. J. ÅSTRÖM and P. EYKHOFF, "System Identification—A Survey," *Automatica 7*, 123-162 (1971).

2. P. EKYHOFF, P. M. VAN DER GRINTEN, H. KWAKERNAAK, and B. P. VELTMAN, "Systems Modeling and Identification," *Proc. IFAC Congr., 3rd, London,* survey paper (1966).

3. T. KAILATH, D. O. MAYNE, and R. K. MEHRA (eds.), "Special Issue on System Identification and Time Series Analysis," *IEEE Trans. Autom. Control AC-19*, 637-951, December 1974.

4. A. LINDQUIST and G. PICCI, "Realization Theory for Multivariate Stationary Gaussian Processes I: State Space Construction," *Int. Symp. Math. Theory Networks Syst., 4th,* 140-148 (1979).

5. A. LINDQUIST and G. PICCI, "Realization Theory for Multivariate Gaussian Processes II: State Space Theory Revisited and Dynamical Representations of Finite Dimensional State Spaces," *Proc. Int. Conf. Inf. Sci. Syst., 2nd, Patras, Greece, 1979*, 108-129 (1980).

6. R. E. KALMAN, "Realization of Covariance Sequences," *Proc. Toeplitz Mem. Conf., Tel Aviv Univ., Israel,* May 12, 1981.

7. J. RISSANEN and T. KAILATH, "Partial Realizations of Random Systems," *Automatica 8*, 389-396 (1972).

8. A. GELB (ed.), "Applied Optimal Estimation," MIT Press, Cambridge, Massachusetts, 1974.

9. L. S. SHIEH and Y. T. TSAY, "Algebra-Geometric Approach for the Model Reduction of Large Scale Multivariable Systems," *Proc. IEEE 131*, Pt. D, No. 1, 23-26, January 1984.

10. T. T. SOONG, "The Use of Moment Matching in Model Reduction of Large Systems," "Large Engineering Systems 2," State Univ. of New York, Buffalo, New York.

11. J. HICKIN and N. K. SINHA, "Model Reduction for Linear Multivariable Systems," *IEEE Trans. Autom. Control AC-25*, No. 6, 1121-1127, December 1980.

12. Y. INOUYE, "Approximation of Multivariable Linear Systems with Impulse Response and Autocorrelation Sequences," *Automatica 19*, No. 3, 265-277 (1983).

13. G. LASTMAN, N. K. SINHA, and P. ROZSA, "On the Selection of States to Be Retained in a Reduced-Order Model," *IEE Proc. 131*, Pt. D, No. 1, 15-22, January 1984.

14. L. R. PUJARA and K. S. RATTAN, "A Frequency Matching Method for Model Reduction of Digital Control Systems," *Int. J. Control 35*, No. 1, 139-148 (1982).

15. B. D. O. ANDERSON, "The Inverse Problem of Stationary Covariance Generation," *J. Stat. Phys. 2*, No. 2 (1969).

16. Y. BARAM, "Realization and Reduction of Markovian Models from Nonstationary Data," *IEEE Trans. Autom. Control AC-26*, No. 6, 1225-1231, December 1981.

17. A. YOUSUFF and R. E. SKELTON, "Covariance Equivalent Realizations with Application to Model Reduction of Large Scale Systems," *in* "Control and Dynamic Systems, Vol. 22 (C. T. Leondes, ed.), Academic Press, New York, 1984.

18. D. A. WAGIE and R. E. SKELTON, "Covariance Equivalent Realizations of Discrete Systems," *Proc. 23rd IEEE Conf. Decision Control, Las Vegas, Nevada*, 12-14, December 1984.

19. R. E. SKELTON and M. L. DELORENZO, "Flexible Space Structure Control Design via Input/Output Cost Assignment," *J. Guidance Control Dyn.*, in press (1986).

20. P. M. MAKILA, T. WESTERLUND, and H. T. TOIVONEN, "Constrained Linear Quadratic Gaussian Control with Process Application," *Automatica 20*, No. 1, 15-29 (1984).

21. J. G. LIN, "Multiple-Objective Problems: Pareto-Optimal Solutions by Method of Proper Equality Constraints," *IEEE Trans. Autom. Control AC-21*, No. 5, October 1976.

22. V. PARETO, "Manual of Political Economy," (A. S. Schwier, trans.), MacMillan, New York, 1971.

23. H. KWAKERNAAK and R. SIVAN, "Linear Optimal Control Systems," Wiley, New York, 1972.

24. R. BARTELS and G. STEWART, "A Solution of the Equation AX + XB = C," *Commun. Assoc. Comput. Mach. 15*, 820-826 (1972).

25. G. GOLUB, S. NASH, and C. VAN LOAN, "A Hessenberg-Schur Method for the Problem AX + XB = C," *IEEE Trans. Autom. Control AC-24*, 909-913 (1979).

26. S. BARNETT and C. STOREY, "Analysis and Synthesis of Stability Matrices," *J. Differ. Equations 3*, 414-422 (1967).

27. T. KAILATH, "Linear Systems," Prentice-Hall, Englewood Cliffs, New Jersey, 1980.

28. J. C. Doyle, G. Stein, "Robustness with Observers," *IEEE Trans. Autom. Control AC-24*, 607-611 (1979).

29. D. GANGSAAS, U. LY, and D. C. NORMAN, "Practical Gust Load Alleviation and Flutter Suppression Control-Laws Based on a LQG Methodology," *Proc. AIAA Aerosp. Sci. Meet. 19th, St. Louis, Missouri*, January 1981.

30. M. ATHANS and P. C. FALB, "Optimal Control," McGraw-Hill, New York, 1966.

31. B. D. O. ANDERSON and J. B. MOORE, "Linear Optimal Control," Prentice-Hall, Englewood Cliffs, New Jersey, 1971.

32. R. V. PATEL and N. MUNRO, "Multivariable System Theory and Design," Pergamon, Oxford, 1982.

33. R. E. KALMAN, "When is a Linear Control System Optimal?" *J. Basic Engr.*, 51-60, March 1964.

34. B. D. O. ANDERSON, "The Inverse Problem of Optimal Control," Stanford Electronics Lab., Technical Report 6560-3, April 1966.

35. B. P. MOLINARI, "The Stable Regulator Problem and Its Inverse," *IEEE Trans. Autom. Control AC-18*, No. 5, October 1973.

36. J. C. WILLEMS, "Least Squares Stationary Optimal Control and the Algebraic Riccati Equation," *IEEE Trans. Autom. Control AC-16*, 621-624, December 1971.

37. A. JAMESON and E. KREINDLER, "Inverse Problem of Linear Optimal Control," *SIAM J. Control 11*, No. 1, February 1973.

38. M. G. SAFONOV and M. ATHANS, "Gain and Phase Margin for Multiloop LQG Regulators," *IEEE Trans. Autom. Control AC-22*, 173-179, April 1977.

39. R. V. PATEL, M. TODA, and B. SINDHAR, "Robustness of Linear Quadratic State Feedback Designs in the Presence of Systems Uncertainty," *IEEE Trans. Autom. Control AC-22*, 945-949, December 1977.

40. E. KREINDLER and A. JAMESON, "Optimality of Linear Control Systems," *IEEE Trans. Autom. Control AC-17*, 349-351 (1972).

41. A. E. BRYSON and Y. C. HO, "Applied Optimal Control," Hemisphere, New York, 1975.

42. E. KREINDLER and A. Jameson, "Conditions for Nonnegativeness of Partitioned Matrices," *IEEE Trans. Autom. Control AC-17*, 147-148 (1972).

43. A. BEN-ISRAEL and T. N. E. GREVILLE, "Generalized Inverses: Theory and Application," Wiley, New York, 1974.

Adaptive Control with Recursive Identification for Stochastic Linear Systems

H. F. CHEN
L. GUO

Institute of Systems Science
Academia Sinica
Beijing, People's Republic of China

I. INTRODUCTION

Parameter-adaptive control and parameter estimation without monitoring in linear, stochastic, discrete-time systems have drawn much attention from control theorists. The self-tuning regulator was first introduced by Åström and Wittenmark [1] and stimulated a great amount of theoretical and practical work in adaptive control. Recently, there appeared a sequence of papers establishing the stabilizing properties of various adaptive control algorithms (see, e.g., [2-5]). At the same time different approaches were introduced for analysis of the parameter estimation of systems without monitoring and various conditions for strong consistency of estimates were proposed (see, e.g., [6-15]). Clearly, since the unknown parameters of the system must usually be estimated for adaptive control, it is natural to ask whether the estimate needed for the system in question is strongly consistent. Unfortunately, the answer is negative in general, as shown in [16]. In order to achieve consistent parameter estimation for adaptive control systems, a randomization technique was used in [17-19] with the result that the system is no longer asymptotically optimal but suboptimal. Not long ago, the authors [20] gave an adaptive control by which both strong consistency of the estimate and optimality of adaptive tracking can be achieved simultaneously. An overview of convergence theory for adaptive control algorithms was presented by Goodwin *et al.* [21].

We establish in Section II some results on strong consistency of parameter estimates for systems without monitoring by not invoking persisten excitation like conditions. Although these results are designed for the later sections, they are interesting by themselves. Section III proves global convergence for adaptive tracking systems and emphasizes the existence of adaptive control law. In Section IV we present the main results of this contribution; namely, we give the optimal adaptive control by which the long-run average of the tracking error reaches its minimum and, at the same time, the parameter estimates are strongly consistent. The last section briefly concerns the convergence rate in both tracking and estimating.

II. STRONG CONSISTENCY OF PARAMETER ESTIMATES FOR SYSTEMS WITHOUT MONITORING

Let $(\Omega, \mathscr{F}, P)$ be a probability space with a family $\{\mathscr{F}_n\}$ of nondecreasing sub-σ-algebras.

Consider the linear stochastic system

$$y_n + A_1 y_{n-1} + \cdots + A_p y_{n-p} = B_1 u_{n-1} + \cdots + B_q u_{n-q} + w_n + C_1 w_{n-1} + \cdots + C_r w_{n-r}, \tag{1}$$

where y_n, u_n, and w_n are the m-, l-, and m-dimensional output, input, and driven noise, respectively, and A_i, B_j, and C_k ($i = 1, \ldots, p$, $j = 1, \ldots, q$, $k = 1, \ldots, r$) are the unknown matrices.

In this section for the system (1) there is no performance index considered, and the control u_n is required to depend on the past measurements only. Throughout the sequel, $\|x\|$ denotes the Euclidean norm of a vector x and $\|A\|$ denotes the maximum singular value of a matrix A.

Let z be the shift-back operator and set

$$A(z) = I + A_1 z + \cdots + A_p z^p, \tag{2}$$

$$B(z) = B_1 + B_2 z + \cdots + B_q z^{q-1}, \tag{3}$$

$$C(z) = I + C_1 z + \cdots + C_r z^r, \tag{4}$$

$$\theta^\tau = [-A_1 \cdots -A_p \quad B_1 \cdots B_q \quad C_1 \cdots C_r]. \tag{5}$$

The task here is to estimate θ recursively at each time n. Denote the nth estimate for θ by θ_n, which can be given by various recursive estimation algorithms. We assume that θ_n is defined by the stochastic gradient algorithm

$$\theta_{n+1} = \theta_n + (\varphi_n/r_n)\left(y_{n+1}^\tau - \varphi_n^\tau \theta_n\right), \tag{6}$$

where

$$\varphi_n^\tau = \Big[y_n^\tau \cdots y_{n-p+1}^\tau \quad u_n^\tau \cdots u_{n-q+1}^\tau \quad y_n^\tau - \varphi_{n-1}^\tau \theta_{n-1} \cdots y_{n-r+1}^\tau - \varphi_{n-r}^\tau \theta_{n-r}\Big], \tag{7}$$

$$r_n = 1 + \sum_{i=1}^{n} \|\varphi_i\|^2, \qquad r_0 = 1. \tag{8}$$

The initial values θ_0 and φ_0 can be arbitrarily chosen.

For the driven noise $\{w_n\}$ we suppose that w_n is $\mathscr{F}_n$ measurable, $w_i = 0$ for $i < 0$, and

$$E(w_n \mid \mathscr{F}_{n-1}) = 0, \tag{9}$$

$$E\left(\|w_n\|^2 \mid \mathscr{F}_{n-1} \right) \le c_0 r_{n-1}^{\epsilon}, \quad \forall \quad n \ge 1, \tag{10}$$

with $c_0 > 0$ and $\epsilon \in [0, 1)$.

It is worth noting that condition (10) is more general than the uniform boundedness condition $E\left(\|w_n\|^2 \mid \mathscr{F}_{n-1} \right) \le \sigma^2$ since $r_n \ge 1$.

To consider the strong consistency of θ_n we introduce matrix $\Phi(n, i)$ defined recursively by

$$\Phi(n + 1, i) = I - \frac{\varphi_n \varphi_n^{\tau}}{r_n} \Phi(n, i), \quad \Phi(i, i) = I. \tag{11}$$

In the sequel we shall see that the properties of $\Phi(n, i)$ are of great importance for parameter estimation.

Set

$$\tilde{\theta}_n = \theta - \theta_n, \tag{12}$$

$$\xi_n = y_n - w_n - \theta_{n-1}^{\tau} \varphi_{n-1}. \tag{13}$$

Lemma 2.1. If $C(z) - \frac{1}{2} I$ is strictly positive real, then

$$\sum_{n=0}^{\infty} \frac{\|\xi_{n+1}\|^2}{r_n} < \infty \quad \text{a.s.}$$

Proof. Noticing that

$$\frac{\|y_i\|^2}{r_n} < 1, \quad \frac{\|u_j\|^2}{r_n} < 1, \quad i = n, \ldots, n - p + 1,$$
$$j = n, \ldots, n - q + 1,$$

$$E\frac{\|w_k\|^2}{r_n} \le E\frac{\|w_k\|^2}{r_{k-1}} = E\left[\frac{1}{r_{k-1}} E\left(\|w_k\|^2 \mid \mathscr{F}_{k-1} \right)\right] \le c_0 E \frac{1}{r_{k-1}^{1-E}} \le c_0,$$

where $k = n + 1, \ldots, n - r + 1$, we have by (1)

$$E\frac{\|y_{n+1}\|^2}{r_n} < \infty.$$

From (6) it follows that

$$E\|\theta_{n+1}\|^2 \le 2E\|\theta_n\|^2 + 4E\left(\frac{\|y_{n+1}\|^2\|\varphi_n\|^2}{r_n^2} + \frac{\|\theta_n\|^2\|\varphi_n\|^2}{r_n^2}\right)$$

$$\le 4E\frac{\|y_{n+1}\|^2}{r_n} + 6E\|\theta_n\|^2.$$

Then by induction we conclude that

$$E\|\theta_n\|^2 < \infty, \quad E\|\tilde{\theta}_n\|^2 < \infty \quad \forall \; n. \tag{14}$$

From (13) it is easy to see that

$$E\frac{1}{r_n}\|\xi_{n+1}\|^2 < \infty \quad \forall \; n; \tag{15}$$

then by (1), (4), (5), and (7) it follows that

$$C(z)\left(y_n - w_n - \theta_{n-1}^{\tau}\varphi_{n-1}\right)$$

$$= \left[(y_n - C(z)w_n) + (C(z) - I)\left(y_n - \theta_{n-1}^{\tau}\varphi_{n-1}\right)\right] - \theta_{n-1}^{\tau}\varphi_{n-1}$$

$$= \theta^{\tau}\varphi_{n-1} - \theta_{n-1}^{\tau}\varphi_{n-1} = \tilde{\theta}_{n-1}^{\tau}\varphi_{n-1}$$

and hence

$$C(z)\xi_n = \tilde{\theta}_{n-1}^{\tau}\varphi_{n-1}. \tag{16}$$

From this and from the strictly positive realness assumption of $C(z) - \frac{1}{2}I$, we know that there exist constants $k_1 > 0$ and $k_2 > 0$ such that

$$s_n \triangleq 2\sum_{i=1}^{n}\xi_i^{\tau}\left(\tilde{\theta}_{i-1}^{\tau}\varphi_{i-1} - \frac{1 + k_1}{2}\xi_i\right) + k_2 \ge 0 \quad \forall \; n. \tag{17}$$

Set

$$M_n \triangleq \operatorname{tr} \tilde{\theta}_n^\tau \tilde{\theta}_n + \frac{S_n}{r_{n-1}} + c_0 E\left[\sum_{i=1}^{\infty} \frac{\|\varphi_i\|^2}{r_i^{2-\epsilon}} \,\middle|\, \mathscr{F}_n\right]$$

$$- c_0 \sum_{i=1}^{n-1} \frac{\|\varphi_i\|^2}{r_i^{2-\epsilon}} + k_1 \sum_{i=1}^{n-1} \frac{\|\xi_{i+1}\|^2}{r_i}. \tag{18}$$

We need the following facts [22]: Let $A_n = 1 + \Sigma_{i=1}^n a_i$, $a_i \geq 0$, $A_0 = 1$; then

$$\sum_{i=1}^{\infty} \frac{a_i}{A_i^\alpha} \leq \frac{1}{\alpha - 1} \quad \forall \ \alpha > 1 \tag{19}$$

and

$$\sum_{i=1}^{\infty} \frac{a_i}{A_i} = \infty \iff A_i \xrightarrow[i\to\infty]{} \infty. \tag{20}$$

By (15), (17), (19), and (10) we conclude that M_n defined by (18) is meaningful and $M_n \geq 0$, $EM_n < \infty$, $\forall\ n \geq 1$.

Since ξ_{n+1} is $\mathscr{F}_n$ measurable and

$$\tilde{\theta}_{n+1} = \tilde{\theta}_n - \frac{\varphi_n}{r_n}\left(\xi_{n+1}^\tau + w_{n+1}^\tau\right),$$

from (18) by taking conditional expectation we have

$$E(M_{n+1} \mid \mathscr{F}_n) \leq \operatorname{tr} \tilde{\theta}_n^\tau \tilde{\theta}_n - \frac{2}{r_n}\left[\xi_{n+1}^\tau\left(\tilde{\theta}_n^\tau \varphi_n - \frac{1 + k_1}{2} \xi_{n+1}\right)\right]$$

$$- \frac{k_1}{r_n}\|\xi_{n+1}\|^2 + \frac{\|\varphi_n\|^2}{r_n^2} E\left(\|w_{n+1}\|^2 \mid \mathscr{F}_n\right) + \frac{S_n}{r_n}$$

$$+ \frac{2\xi_{n+1}^\tau\left\{\tilde{\theta}_n^\tau \varphi_n - [(1 + k_1)/2]\xi_{n+1}\right\}}{r_n}$$

$$+ c_0 E\left[\sum_{i=1}^{\infty} \frac{\|\varphi_i\|^2}{r_i^{2-\epsilon}} \mid \mathscr{F}_n\right] - c_0 \sum_{i=1}^{n} \frac{\|\varphi_i\|^2}{r_i^{2-\epsilon}} + k_1 \sum_{i=1}^{n} \frac{\|\xi_{i+1}\|^2}{r_i}$$

$$\leq M_n.$$

Hence $(M_n, \mathscr{F}_n)$ is a nonegative supermartingale and by the convergence theorem we have

$$M_n \xrightarrow[n\to\infty]{} M < \infty \quad \text{a.s.}$$

From here and (18) the assertion of the lemma follows immediately. ∎

Lemma 2.2. The following estimate takes place for any $n \geq 0$:

$$\sum_{i=0}^{n-1} \frac{\|\Phi(n, i + 1)\varphi_i\|^2}{r_i} \leq d,$$

where $d = mp + lq + mr$.

Proof. The desired result is verified by the following chain of inequalities:

$$\begin{aligned}
d &= \operatorname{tr} \Phi(n, n)\Phi^\tau(n, n) \\
&\geq \sum_{i=0}^{n-1} \operatorname{tr}[\Phi(n, i + 1)\Phi^\tau(n, i + 1) - \Phi(n, i)\Phi^\tau(n, i)] \\
&= \operatorname{tr} \sum_{i=0}^{n-1} \Phi(n, i + 1)[I - \Phi(i + 1, i)\Phi^\tau(i + 1, i)]\Phi^\tau(n, i + 1) \\
&= \operatorname{tr} \sum_{i=0}^{n-1} \Phi(n, i + 1)\left[I - \left(I - \frac{\varphi_i\varphi_i^\tau}{r_i}\right)\left(I - \frac{\varphi_i\varphi_i^\tau}{r_i}\right)\right]\Phi^\tau(n, i + 1) \\
&= \operatorname{tr}\sum_{i=0}^{n-1} \Phi(n, i + 1)\left[\frac{\varphi_i\varphi_i^\tau}{r_i} + \frac{1}{r_i}\varphi_i\left(I - \frac{\|\varphi_i\|^2}{r_i} I\right)\varphi_i^\tau\right]\Phi^\tau(n, i + 1) \\
&\geq \operatorname{tr} \sum_{i=0}^{n-1} \Phi(n, i + 1) \frac{\varphi_i\varphi_i^\tau}{r_i} \Phi^\tau(n, i + 1) \\
&= \sum_{i=0}^{n-1} \frac{\|\Phi(n, i + 1)\varphi_i\|^2}{r_i}. \quad \blacksquare
\end{aligned}$$

Theorem 2.1

Assume the transfer matrix $C(z) - \frac{1}{2}I$ is strictly positive real. If $\Phi(n, 0) \xrightarrow[n\to\infty]{} 0$ then $\theta_n \xrightarrow[n\to\infty]{} \theta$ a.s. for any initial value θ_0. For the special case $r = 0$, the converse assertion is also true; that is, if $\theta_n \xrightarrow[n\to\infty]{} \theta$ a.s. for any θ_0, then $\Phi(n, 0) \xrightarrow[n\to\infty]{} 0$ a.s.

Proof. Set

$$\varphi_n^0 = \left[y_n^\tau \cdots y_{n-p+1}^\tau \quad u_n^\tau \cdots u_{n-q+1}^\tau \quad w_n^\tau \cdots w_{n-r+1}^\tau\right]^\tau, \tag{21}$$

$$\varphi_n^\xi = \left[0 \cdots 0 \quad 0 \cdots 0 \quad \xi_n^\tau \cdots \xi_{n-r+1}^\tau\right]^\tau. \tag{22}$$

Clearly, we have

$$\varphi_n = \varphi_n^0 + \varphi_n^\xi, \tag{23}$$

with

$$\sum_{n=0}^{\infty} \frac{\|\varphi_n^\xi\|^2}{r_n} < \infty \quad \text{a.s.} \tag{24}$$

by Lemma 2.1.

From (1), (5), (21), y_{n+1} can be written in the form

$$y_{n+1} = \theta^\tau \varphi_n^0 + w_{n+1};$$

hence we obtain

$$\theta_{n+1} = \theta_n + \frac{\varphi_n}{r_n}\left(\varphi_n^{0\tau}\theta + w_{n+1}^\tau - \varphi_n^\tau\theta_n\right)$$

$$= \theta_n + \frac{\varphi_n}{r_n}\left(\varphi_n^\tau\theta - \varphi_n^{\xi\tau}\theta + w_{n+1}^\tau - \varphi_n^\tau\theta_n\right),$$

$$= \theta_n + \frac{\varphi_n}{r_n}\left(\varphi_n^\tau\tilde{\theta}_n - \varphi_n^{\xi\tau}\theta + w_{n+1}^\tau\right),$$

$$\tilde{\theta}_{n+1} = \left(I - \frac{\varphi_n\varphi_n^\tau}{r_n}\right)\tilde{\theta}_n + \frac{\varphi_n\varphi_n^{\xi\tau}}{r_n}\theta - \frac{\varphi_n}{r_n}w_{n+1}^\tau,$$

and finally

$$\tilde{\theta}_{n+1} = \Phi(n + 1, 0)\tilde{\theta}_0 + \sum_{j=0}^{n} \Phi(n + 1, j + 1)\frac{\varphi_j\varphi_j^{\xi\tau}}{r_j}\theta$$

$$- \sum_{j=0}^{n} \Phi(n + 1, j + 1)\frac{\varphi_j}{r_j} w_{j+1}^{\tau}. \tag{25}$$

Now assume $\Phi(n, 0) \xrightarrow[n\to\infty]{} 0$. By (24), Lemma 2.2, and the Schwarz inequality, it follows that

$$\left\|\sum_{j=0}^{n} \Phi(n + 1, j + 1)\frac{\varphi_j\varphi_j^{\xi\tau}}{r_j}\right\|$$

$$\leq \left\|\sum_{j=0}^{N} \Phi(n + 1, j + 1)\frac{\varphi_j\varphi_j^{\xi\tau}}{r_j}\right\|$$

$$+ \left(\sum_{j=N+1}^{n} \frac{\|\Phi(n + 1, j + 1)\varphi_j\|^2}{r_j}\right)^{1/2}\left(\sum_{j=N+1}^{n} \frac{\|\varphi_j^{\xi}\|^2}{r_j}\right)^{1/2}$$

$$\leq \left\|\sum_{j=0}^{N} \Phi(n + 1, j + 1)\frac{\varphi_j\varphi_j^{\xi\tau}}{r_j}\right\|$$

$$+ \sqrt{d}\left(\sum_{j=N+1}^{n} \frac{\|\varphi_j^{\xi}\|^2}{r_j}\right)^{1/2} \xrightarrow[\substack{n\to\infty \\ N\to\infty}]{} 0. \tag{26}$$

Thus the first two terms on the right-hand side of (25) vanish as $n \to \infty$, and so the main task is to consider its last term.

We note at once that by the martingale convergence theorem [23] that

$$\sum_{j=1}^{\infty} \frac{\varphi_j w_{j+1}^{\tau}}{r_j^{1-\delta}}$$

converges a.s. for

$$\forall\ \delta \in \left[0,\ \frac{1-\epsilon}{2}\right)$$

since by (10) and (19)

$$\sum_{i=1}^{\infty} E\left[\left\|\frac{\varphi_i w_{i+1}^{\tau}}{r_i^{1-\delta}}\right\|^2 \middle| \mathscr{F}_i\right] \le c_0 \sum_{i=1}^{\infty} \frac{\|\varphi_i\|^2}{r_i^{2-\epsilon-2\delta}} < \infty \quad \text{a.s.}$$

Then we have

$$S_n^{\delta} \triangleq \sum_{i=n}^{\infty} \frac{\varphi_i w_{i+1}^{\tau}}{r_i^{1-\delta}} = o(1) \quad \text{as} \quad n \to \infty$$

and

$$\begin{aligned}
\left\| r_n^{\delta} \sum_{i=n}^{\infty} \frac{\varphi_i w_{i+1}^{\tau}}{r_i} \right\| &= \left\| r_n^{\delta} \sum_{i=n}^{\infty} \frac{\varphi_i w_{i+1}^{\tau}}{r_i^{1-\delta}} \cdot \frac{1}{r_i^{\delta}} \right\| \\
&= \left\| r_n^{\delta} \sum_{i=n}^{\infty} \left(S_i^{\delta} - S_{i+1}^{\delta}\right) \cdot \frac{1}{r_i^{\delta}} \right\| \\
&= \left\| S_n^{\delta} - r_n^{\delta} \sum_{i=n}^{\infty} S_{i+1}^{\delta} \left(\frac{1}{r_i^{\delta}} - \frac{1}{r_{i+1}^{\delta}}\right) \right\| \\
&= o(1) + o(1) \sum_{i=n}^{\infty} r_n^{\delta} \left(\frac{1}{r_i^{\delta}} - \frac{1}{r_{i+1}^{\delta}}\right) \\
&= o(1), \quad n \to \infty \qquad (27)
\end{aligned}$$

From this we conclude that $\tilde{S}_n \to 0$ and that there exists $c > 0$ possibly depending on ω such that $\|\tilde{S}_{n-1}\| \le c r_n^{-\delta}$, where by definition we have set

$$\tilde{S}_n = S - S_n, \quad S_n = \sum_{i=0}^{n} \frac{\varphi_i}{r_i} w_{i+1}^{\tau}, \quad S = \sum_{i=0}^{\infty} \frac{\varphi_i}{r_i} w_{i+1}^{\tau}, \quad S_{-1} = 0.$$

We now estimate the last term of (25) as follows

$$\left\| \sum_{j=0}^{n} \Phi(n+1, j+1) \frac{\varphi_j}{r_j} w_{j+1}^{\tau} \right\|$$

$$= \left\| \sum_{j=0}^{n} \Phi(n+1, j+1)(S_j - S_{j-1}) \right\|$$

$$= \left\| S_n - \sum_{j=0}^{n} [\Phi(n+1, j+1) - \Phi(n+1, j)] S_{j-1} \right\|$$

$$= \left\| S_n - \sum_{j=0}^{n} [\Phi(n+1, j+1) - \Phi(n+1, j)] S \right.$$

$$\left. + \sum_{j=0}^{n} [\Phi(n+1, j+1) - \Phi(n+1, j)] \tilde{S}_{j-1} \right\|$$

$$= \left\| S_n - S + \Phi(n+1, 0) S \right.$$

$$\left. + \sum_{j=0}^{n} \Phi(n+1, j+1)[I - \Phi(j+1, j)] \tilde{S}_{j-1} \right\|$$

$$\leq \|\tilde{S}_n\| + \|\Phi(n+1, 0) S\|$$

$$+ \sum_{j=0}^{n} \left\| \Phi(n+1, j+1) \frac{\varphi_j \varphi_j^{\tau}}{r_j} \tilde{S}_{j-1} \right\|$$

$$\leq \|\tilde{S}_n\| + \|\Phi(n+1, 0) S\|$$

$$+ c \sum_{j=0}^{n} \frac{\|\Phi(n+1, j+1)\varphi_j\|}{r_j^{1/2}} \cdot \frac{\|\varphi_j\|}{r_j^{1/2+\delta}}$$

$$\leq \|\tilde{S}_n\| + \|\Phi(n + 1, 0)S\|$$

$$+ c \sum_{j=0}^{N} \frac{\|\Phi(n + 1, j + 1)\varphi_j\|}{r_j^{1/2}} \cdot \frac{\|\varphi_j\|}{r_j^{1/2+\delta}}$$

$$+ c\left(\sum_{j=N+1}^{n} \frac{\|\Phi(n + 1, j + 1)\varphi_j\|^2}{r_j}\right)^{1/2}\left(\sum_{j=N+1}^{n} \frac{\|\varphi_j\|^2}{r_j^{1+2\delta}}\right)^{1/2}.$$

On the right-hand side of the preceding expression the first three terms go to 0 as $n \to \infty$ for any N, while the last term tends to zero as $N \to \infty$ by (19) and Lemma 2.2.

Thus we have shown that $\Phi(n, 0) \xrightarrow[n\to\infty]{} 0$ implies $\theta_n \to \theta$.

For the special case $r = 0$ the expression (25) becomes

$$\tilde{\theta}_{n+1} = \Phi(n + 1, 0)\tilde{\theta}_0 - \sum_{j=0}^{n} \Phi(n + 1, j + 1)\frac{\varphi_j}{r_j} w_{j+1}^{\tau}. \tag{28}$$

We note that the last term is independent of θ_0; then if $\tilde{\theta}_n \to 0$ for any initial value θ_0, it necessarily follows that $\Phi(n + 1, 0)\tilde{\theta}_0 \xrightarrow[n\to\infty]{} 0$, for any θ_0. Hence we have $\Phi(n + 1, 0) \xrightarrow[n\to\infty]{} 0$. ■

Theorem 2.1 tells us that $\Phi(n, 0) \xrightarrow[n\to\infty]{} 0$ is a key condition guaranteeing strong consistency of the estimate. We now compare it with the well-known persistent excitation condition indexed by (a) or (b) below, which are usually assumed for strong consistency of the estimate given by recursive algorithms (e.g., [6], [8], [13]):

(a) $\frac{1}{n} \sum_{j=1}^{n} \varphi_j \varphi_j^{\tau} \xrightarrow[n\to\infty]{} R > 0$ a.s.

(b) $r_n \to \infty$ and $\lambda_{max}^n/\lambda_{min}^n \leq \gamma < \infty$, $\forall\, n \geq 0$, a.s., where λ_{max}^{n} and λ_{min}^{n}, respectively, denote the maximum and minimum eigenvalue of matrix $\Sigma_{j=1}^{n} \varphi_j \varphi_j^{\tau} + (1/d)I$, with d being the dimension of φ_n and γ may depend on ω.

Obviously, condition (a) implies condition (b) and the persistent excitation condition means that the matrix $\Sigma_{j=0}^{n} \varphi_j \varphi_j^{\tau}$ is not ill conditioned. In Theorem 2.2, which follows, we shall see that it is still possible that the estimate is consistent even though the matrix $\Sigma_{j=0}^{n} \varphi_j \varphi_j^{\tau}$ is ill conditioned.

Theorem 2.2

If $r_n \to \infty$,

$$\overline{\lim_{n\to\infty}}\, r_n/r_{n-1} < \infty$$

and there exist quantities N_0 and M possibly depending on ω such that

$$\lambda_{max}^{n}/\lambda_{min}^{n} \leq M(\log r_n)^{1/4} \quad \text{a.s.} \quad \forall \quad n \geq N_0,$$

then

$$\Phi(n, 0) \xrightarrow[n\to\infty]{} 0 \quad \text{a.s.}$$

We first prove lemmas. Let

$$m(t) \triangleq \max[n : t_n \leq t], \quad t \geq 0, \tag{29}$$

$$t_n \triangleq \sum_{i=2}^{n-1} \frac{\| \varphi_i \|^2}{r_i (\log r_{i-1})^{1/4}}.$$

Lemma 2.3. Under the conditions of Theorem 2.2, there are positive quantities α, β, and N, which are possibly depending on ω such that

$$\sum_{i=m(N+(k-1)\alpha)}^{m(N+k\alpha)-1} \frac{\varphi_i \varphi_i^{\tau}}{r_i} \geq \beta I \quad \text{a.s.} \quad \forall \quad k \geq 1. \tag{30}$$

Proof. We first show for any t

$$m(t) < \infty, \quad \forall \ t. \tag{31}$$

From the condition

$$\overline{\lim_{n\to\infty}} r_n/r_{n-1} < \infty,$$

it is clear that there is a positive and possibly depending on ω, $l \in (0, \infty)$ such that

$$r_n/r_{n-1} \le l, \quad \forall \ n \ge 1. \tag{32}$$

Then we have

$$t_n = \sum_{i=2}^{n-1} \frac{\|\varphi_i\|^2}{r_i(\log r_{i-1})^{1/4}} \ge \frac{1}{l} \sum_{i=2}^{n-1} \frac{\|\varphi_i\|^2}{r_{i-1}(\log r_{i-1})^{1/4}}$$

$$= \frac{1}{l} \sum_{i=2}^{n-1} \int_{r_{i-1}}^{r_i} \frac{dt}{r_{i-1}(\log r_{i-1})^{1/4}}$$

$$\ge \frac{1}{l} \sum_{i=2}^{n-1} \int_{r_{i-1}}^{r_i} \frac{dt}{t(\log t)^{1/4}} = \frac{1}{l} \int_{r_1}^{r_{n-1}} \frac{dt}{t(\log t)^{1/4}}$$

$$= \frac{4}{3l}\left[\log^{3/4} r_{n-1} - \log^{3/4} r_1\right]. \tag{33}$$

From here it follows that $t_n \to \infty$ by $r_n \to \infty$, and then (31) is verified by the definition (29). By (31) there exists N such that $m(N) \ge N_0$ and

$$(\log r_i)^{1/4} \ge 1, \qquad \frac{(\log r_i)^{1/4}}{r_i} \le \frac{1}{2M}, \qquad \forall \ i \ge m(N). \tag{34}$$

By summation by parts for any $k \ge 1$ we obtain

$$\sum_{i=m(N+(k-1)\alpha)}^{m(N+k\alpha)-1} \frac{\varphi_i\varphi_i^\tau}{r_i} \ge \sum_{i=m(N+(k-1)\alpha)}^{m(N+k\alpha)} \frac{\varphi_i\varphi_i^\tau}{r_i} - I$$

$$= \sum_{i=m(N+(k-1)\alpha)}^{m(N+k\alpha)} \frac{1}{r_i}\left(\sum_{j=1}^{i} \varphi_j\varphi_j^\tau - \sum_{j=1}^{i-1} \varphi_j\varphi_j^\tau\right) - I$$

$$= \frac{1}{r_{m(N+k\alpha)}} \sum_{j=1}^{m(N+k\alpha)} \varphi_j \varphi_j^\tau - \frac{1}{r_{m(N+(k-1)\alpha)}} \sum_{j=1}^{m(N+(k-1)\alpha)-1} \varphi_j \varphi_j^\tau$$

$$+ \sum_{i=m(N+(k-1)\alpha)+1}^{m(N+k\alpha)} \sum_{j=1}^{i-1} \varphi_j \varphi_j^\tau \left(\frac{1}{r_{i-1}} - \frac{1}{r_i} \right) - I$$

$$\geq \sum_{i=m(N+(k-1)\alpha)+1}^{m(N+k\alpha)} \sum_{j=1}^{i-1} \varphi_j \varphi_j^\tau \frac{\|\varphi_i\|^2}{r_{i-1} \cdot r_i} - 2I$$

$$\geq \sum_{i=m(N+(k-1)\alpha)+1}^{m(N+k\alpha)} \left(\lambda_{\min}^{i-1} - \frac{1}{d} \right) I \frac{\|\varphi_i\|^2}{r_{i-1} \cdot r_i} - 2I$$

$$\geq \sum_{i=m(N+(k-1)\alpha)+1}^{m(N+k\alpha)} \left(\frac{\lambda_{\max}^{i-1}}{M(\log r_{i-1})^{1/4}} - \frac{1}{d} \right) I \frac{\|\varphi_i\|^2}{r_{i-1} \cdot r_i} - 2I$$

$$\geq \frac{1}{d} \cdot \sum_{i=m(N+(k-1)\alpha)+1}^{m(N+k\alpha)} \left(\frac{r_{i-1}}{M(\log r_{i-1})^{1/4}} - 1 \right) I \frac{\|\varphi_i\|^2}{r_{i-1} \cdot r_i} - 2I$$

$$= \frac{1}{d} \sum_{i=m(N+(k-1)\alpha)+1}^{m(N+k\alpha)} \left(\frac{1}{M} - \frac{(\log r_{i-1})^{1/4}}{r_{i-1}} \right) \cdot \frac{\|\varphi_i\|^2}{r_i (\log r_{i-1})^{1/4}} - 2I$$

$$\geq \frac{1}{2Md} (t_{m(N+k\alpha)+1} - t_{m(N+(k-1)\alpha)+1}) I - 2I$$

$$\geq \frac{1}{2Md} [N + k\alpha - (N + (k-1)\alpha + 1)] I - 2I$$

$$= \left[\frac{1}{2Md} (\alpha - 1) - 2 \right] I.$$

The assertion of the lemma follows by taking $\alpha > 4Md + 1$, $\beta \triangleq (1/2Md)(\alpha - 1) - 2$. ∎

Lemma 2.4. Under the conditions of Theorem 2.2 there exists a positive c_1 independent of k such that

$$\|\Phi(m(N + k\alpha),\ m(N + (k - 1)\ ))\| \leq \sqrt{1 - \frac{\beta^2}{c_1 k}}, \qquad \forall \ \ k \geq 1,$$

where β is given by Lemma 2.3.

Proof. Let ρ_k be the maximum eigenvalue of the matrix

$$\Phi^\tau(m(N + k\alpha),\ m(N + (k - 1)\alpha)) \cdot \Phi(m(N + k\alpha),\ m(N + (k - 1)\alpha))$$

and let $x_{m(N+(k-1)\alpha)}$ be the corresponding unit eigenvector.

By definition of the norm we have

$$\|\Phi(m(N + k\alpha),\ m(N + (k - 1)\alpha))\| = \sqrt{\rho_k}. \tag{35}$$

For $i \in [m(N + (k - 1)\alpha),\ m(N + k\alpha) - 1]$ recursively define vector x_i:

$$x_{i+1} = \left(I - \frac{\varphi_i \varphi_i^\tau}{r_i}\right)x_i. \tag{36}$$

By (11) we have

$$x_{m(N+k\alpha)} = \Phi(m(N + k\alpha),\ m(N + (k - 1)\alpha))x_{m(N+(k-1)\alpha)}.$$

Hence

$$\begin{aligned} x^\tau_{m(N+k\alpha)}x_{m(N+k\alpha)} &= x^\tau_{m(N+(k-1)\alpha)}\Phi^\tau(m(N + k\alpha),\ m(N + (k - 1)\alpha)) \\ &\quad \cdot \Phi(m(N + k\alpha),\ m(N + (k - 1)\alpha))x_{m(N+(k-1)\alpha)} \\ &= x^\tau_{m(N+(k-1)\alpha)} \cdot \rho_k \cdot x_{m(N+(k-1)\alpha)} \\ &= \rho_k. \end{aligned} \tag{37}$$

By use of (36) we obtain

$$x^\tau_{i+1}x_{i+1} = x^\tau_i\left(I - \frac{\varphi_i \varphi_i^\tau}{r_i}\right)\left(I - \frac{\varphi_i \varphi_i^\tau}{r_i}\right)x_i$$

$$= x_i^\tau x_i - x_i^\tau \frac{\varphi_i \varphi_i^\tau}{r_i} x_i - x_i^\tau \left(\frac{\varphi_i \varphi_i^\tau}{r_i} - \frac{\varphi_i \|\varphi_i\|^2 \varphi_i^\tau}{r_i^2} \right) x_i$$

$$\leq x_i^\tau x_i - x_i^\tau \frac{\varphi_i \varphi_i^\tau}{r_i} x_i. \tag{38}$$

Summing up both sides of (38) and using (37) we see that

$$\sum_{i=m(N+(k-1)\alpha)}^{m(N+k\alpha)-1} \frac{\|\varphi_i^\tau x_i\|}{r_i} \leq x_{m(N+(k-1)\alpha)}^\tau x_{m(N+(k-1)\alpha)} - x_{m(N+k\alpha)}^\tau x_{m(N+k\alpha)}$$

$$= 1 - \rho_k. \tag{39}$$

For $i \in [m(N + (k - 1)\alpha), m(N + k\alpha) - 1]$, from (36) we have

$$\|x_i - x_{m(N+(k-1)\alpha)}\| = \left\| \sum_{j=m(N+(k-1)\alpha)}^{i-1} \frac{\varphi_j \varphi_j^\tau}{r_j} x_j \right\|$$

$$\leq \{\log r_{m(N+k\alpha)-1}\}^{1/8} \sum_{j=m(N+(k-1)\alpha)}^{m(N+k\alpha)-1} \frac{\|\varphi_j\|}{r_j^{1/2}\{\log r_{j-1}\}^{1/8}} \cdot \frac{\|\varphi_j^\tau x_j\|}{r_j^{1/2}}$$

$$\leq \{\log r_{m(N+k\alpha)-1}\}^{1/8} \cdot \left\{ \sum_{j=m(N+(k-1)\alpha)}^{m(N+k\alpha)-1} \frac{\|\varphi_j\|^2}{r_j (\log r_{j-1})^{1/4}} \right\}^{1/2} \cdot \left\{ \sum_{j=m(N+(k-1)\alpha)}^{m(N+k\alpha)-1} \frac{\|\varphi_j^\tau x_j\|^2}{r_j} \right\}^{1/2}$$

$$\leq \{\log r_{m(N+k\alpha)-1}\}^{1/8} \sqrt{\alpha + 1} \cdot \sqrt{1 - \rho_k}. \tag{40}$$

where for the last inequality (29) and (39) are invoked.

Finally, by Lemma 2.3 and (39), (40) we can estimate as follows:

$$\beta \le x^{\tau}_{m(N+(k-1)\alpha)} \sum_{i=m(N+(k-1)\alpha)}^{m(N+k\alpha)-1} \frac{\varphi_i\varphi_i^{\tau}}{r_i} x_{m(N+(k-1)\alpha)}$$

$$\le \left\| x^{\tau}_{m(N+(k-1)\alpha)} \sum_{i=m(N+(k-1)\alpha)}^{m(N+k\alpha)-1} \frac{\varphi_i\varphi_i^{\tau}}{r_i}(x_{m(N+(k-1)\alpha)} - x_i)\right\|$$

$$+ \left\| x^{\tau}_{m(N+(k-1)\alpha)} \cdot \sum_{i=m(N+(k-1)\alpha)}^{m(N+k\alpha)-1} \frac{\varphi_i\varphi_i^{\tau}}{r_i} x_i \right\|$$

$$\le \{\log r_{m(N+k\alpha)-1}\}^{1/4} \sum_{i=m(N+(k-1)\alpha)}^{m(N+k\alpha)-1} \frac{\|\varphi_i\|^2}{r_i(\log r_{i-1})^{1/4}}$$

$$\times \|x_{m(N+(k-1)\alpha)} - x_i\|$$

$$+ \{\log r_{m(N+k\alpha)-1}\}^{1/8} \sum_{i=m(N+(k-1)\alpha)}^{m(N+k\alpha)-1} \frac{\|\varphi_i\|}{r_i^{1/2}\cdot(\log r_{i-1})^{1/8}}$$

$$\cdot \frac{\|\varphi_i^{\tau}x_i\|}{r_i^{1/2}}$$

$$\le \{\log r_{m(N+k\alpha)-1}\}^{1/4}(1+\alpha)\{\log r_{m(N+k\alpha)-1}\}^{1/8}\sqrt{\alpha+1}\ \sqrt{1-\rho_k}$$

$$+ \{\log r_{m(N+k\alpha)-1}\}^{1/8} \cdot \sqrt{1+\alpha}\ \sqrt{1-\rho_k}$$

$$= \Big\{(\log r_{m(N+k\alpha)-1})^{3/8}(\alpha+1)^{3/2}$$

$$+ (\log r_{m(N+k\alpha)-1})^{1/8}(\alpha+1)^{1/2}\Big\}\sqrt{1-\rho_k}\,. \qquad (41)$$

But by (29) and (33) we know that

$$t \ge t_{m(t)} \ge \frac{4}{3l}\left[\log^{3/4} r_{m(t)-1} - \log^{3/4} r_1\right], \qquad (42)$$

and hence that

$$\log^{3/8} r_{m(N+k\alpha)-1} \leq \left[\frac{3l}{4}(N + k\alpha) + \log^{3/4} r_1\right]^{1/2} \tag{43}$$

$$\log^{1/8} r_{m(N+k\alpha)-1} \leq \left[\frac{3l}{4}(N + k\alpha) + \log^{3/4} r_1\right]^{1/6}. \tag{44}$$

Then by (41), (43), and (44) we conclude that

$$\beta \leq \left\{(\alpha + 1)^{3/2} + (\alpha + 1)^{1/2}\left[\frac{3l}{4}(N + k\alpha) + \log^{3/4} r_1\right]^{-1/3}\right\}$$
$$\cdot \left\{\frac{3l\alpha}{4} k + \frac{3l}{4} \cdot N + \log^{3/4} r_1\right\}^{1/2} \cdot \sqrt{1 - \rho_k}$$

from which it is obvious that there exists $c_1 > 0$ such that $\beta \leq (c_1 k)^{1/2}(1 - \rho_k)^{1/2}$, $\forall\ k \geq 1$; that is,

$$\rho_k \leq 1 - \frac{\beta^2}{c_1 k} \quad \forall\ k \geq 1.$$

This completes the proof of the lemma. ∎

Proof of Theorem 2.2. From (11) it is easy to see that $\|\Phi(n, 0)\|$ is nonincreasing as $n \to \infty$; then it goes to a finite limit l_0, that is

$$\|\Phi(n, 0)\| \xrightarrow[n\to\infty]{} l_0 \geq 0. \tag{45}$$

By Lemma 2.4 we have

$$\|\Phi(m(N + k\alpha), 0)\| \leq \prod_{i=k}^{k} \|\Phi(m(N + i\alpha), m(N(i - 1)\alpha))\|$$
$$\cdot \|\Phi(m(N), 0)\|$$
$$\leq \left\{\prod_{i=1}^{k}\left(1 - \frac{\beta^2}{c_1 i}\right)\right\}^{1/2} \xrightarrow[k\to\infty]{} 0. \tag{46}$$

since $\Sigma_{i=1}^{\infty} \beta^2/c_1 i = \infty$.

From (45), (46) it is clear that $\Phi(n, 0) \xrightarrow[n\to\infty]{} 0$ since $m(N + k\alpha) \xrightarrow[k\to\infty]{} \infty$. ∎

Remark 2.1. It is worth remarking that for any vector φ_n not necessarily defined by (7) Theorem 2.2 holds true, with r_n and $\Phi(n, 0)$ given by (8) and (11), respectively.

From the definition (7) we see that φ_n and hence $\Phi(n + 1, 0)$ depend on the past estimates θ_i, $i \leq n - 1$. We now give a condition that is equivalent to $\Phi(n, 0) \xrightarrow[n\to\infty]{} 0$ and is independent of estimates θ_i.

Define $\Phi^0(n, i)$ by the recursive equation

$$\Phi^0(n + 1, i) = \left(I - \frac{\varphi_n^0 \varphi_n^{0\tau}}{r_n^0}\right)\Phi^0(n, i), \quad \Phi^0(i, i) = I, \tag{47}$$

where φ_n^0 is given by (21) and

$$r_n^0 = 1 + \sum_{i=1}^{n} \|\varphi_i^0\|^2, \quad r_0^0 = 1. \tag{48}$$

Clearly, $\Phi^0(n, 0)$ is free of θ_i.

Theorem 2.3

Assume that $C(z) - \frac{1}{2} I$ is strictly positive real. Then

$$\Phi(n, 0) \xrightarrow[n\to\infty]{} 0 \quad \text{if and only if} \quad \Phi^0(n, 0) \xrightarrow[n\to\infty]{} 0.$$

Proof. Without loss of generality we assume that $\|\varphi_0\| \neq 1$.

Suppose that $\Phi(n, 0) \to 0$; then from the following chain of equalities

$$\det \Phi(n + 1, 0) = \det \prod_{i=0}^{n} \Phi(i + 1, i) = \prod_{i=0}^{n} \det\left(I - \frac{\varphi_i \varphi_i^\tau}{r_i}\right)$$

$$= \prod_{i=1}^{n} \frac{r_{i-1}}{r_i}\left(1 - \|\varphi_0\|^2\right) = \frac{1}{r_n}\left(1 - \|\varphi_0\|^2\right), \tag{49}$$

we see that $r_n \to \infty$.

By (23), (24), (48), and the Kronecker lemma we have

$$\frac{r_n^0}{r_n} = \frac{r_n - 2\,\Sigma_{i=1}^n \varphi_i^\tau \varphi_i^\xi + \Sigma_{i=1}^n \|\varphi_i^\xi\|^2}{r_n} \xrightarrow[n\to\infty]{} 1. \tag{50}$$

Hence (24) is valid with r_i replaced by r_i^0:

$$\sum_{i=0}^{\infty} \frac{\|\varphi_i^\xi\|^2}{r_i^0} < \infty. \tag{51}$$

We note that

$$\Phi^0(n+1, 0) = \left(I - \frac{\varphi_n \varphi_n^\tau}{r_n}\right)\Phi^0(n, 0) + \left(\frac{\varphi_n \varphi_n^\tau}{r_n} - \frac{\varphi_n^0 \varphi_n^{0\tau}}{r_n^0}\right)\Phi^0(n, 0); \tag{52}$$

then

$$\begin{aligned}\Phi^0(n+1, 0) &= \Phi(n+1, 0) + \sum_{j=0}^{n} \Phi(n+1, j+1) \\ &\quad \times \left(\frac{\varphi_j \varphi_j^\tau}{r_j} - \frac{\varphi_j^0 \varphi_j^{0\tau}}{r_j^0}\right)\Phi^0(j, 0) \\ &= \Phi(n+1, 0) + \sum_{j=0}^{n} \Phi(n+1, j+1)\frac{\varphi_j \varphi_j^{\xi\tau}}{r_j}\Phi^0(j, 0) \\ &\quad + \sum_{j=0}^{n} \Phi(n+1, j+1)\frac{\varphi_j^\xi \cdot \varphi_j^{0\tau}}{r_j^0}\Phi^0(j, 0) \\ &\quad + \sum_{j=0}^{n} \frac{\Phi(n+1, j+1)\varphi_j}{r_j^{1/2}} \\ &\quad \times \left(\sqrt{\frac{r_j^0}{r_j}} - \sqrt{\frac{r_j}{r_j^0}}\right)\frac{\varphi_j^{0\tau}\Phi^0(j, 0)}{r_j^{0\,1/2}}.\end{aligned} \tag{53}$$

By (24), Lemma 2.2, and the fact that $\Phi(n, 0) \underset{n\to\infty}{\longrightarrow} 0$, $\|\Phi^0(n, 0)\| \leq 1$ for the second term on the right-hand side of (53), we estimate as follows:

$$\left\| \sum_{j=0}^{n} \Phi(n + 1, j + 1) \frac{\varphi_j \varphi_j^{\xi\tau}}{r_j} \Phi^0(j, 0) \right\|$$
$$\leq \left\| \sum_{j=0}^{N} \Phi(n + 1, j + 1) \frac{\varphi_j \varphi_j^{\xi\tau}}{r_j} \Phi^0(j, 0) \right\| + \left(\sum_{j=N+1}^{n} \frac{\|\Phi(n + 1, j + 1)\varphi_j\|^2}{r_j} \right)^{1/2} \times \left(\sum_{j=N+1}^{n} \frac{\|\varphi_j^\xi\|^2}{r_j} \right)^{1/2} \underset{\substack{n\to\infty \\ N\to\infty}}{\longrightarrow} 0. \qquad (54)$$

To estimate the third term we first note that

$$d \geq \operatorname{tr} \Phi^{0^\tau}(N, 0)\Phi^0(N, 0)$$
$$\geq \sum_{j=N}^{\infty} \operatorname{tr}[\Phi^{0^\tau}(j, 0)\Phi^0(j, 0) - \Phi^{0^\tau}(j + 1, 0)\Phi^0(j + 1, 0)]$$
$$= \sum_{j=N}^{\infty} \operatorname{tr} \Phi^{0^\tau}(j, 0)\left[I - \left(I - \frac{\varphi_j^0 \varphi_j^{0\tau}}{r_j^0} \right)\left(I - \frac{\varphi_j^0 \varphi_j^{0\tau}}{r_j^0} \right) \right]\Phi^0(j, 0)$$
$$\geq \sum_{j=N}^{\infty} \frac{\|\Phi^{0^\tau}(j, 0)\varphi_j^0\|^2}{r_j^0}; \qquad (55)$$

then by (51) and (55) we obtain

$$\left\| \sum_{j=0}^{n} \Phi(n + 1, j + 1) \frac{\varphi_j^\xi \varphi_j^{0\tau}}{r_j^0} \Phi^0(j, 0) \right\|$$
$$\leq \left\| \sum_{j=0}^{N} \Phi(n + 1, j + 1) \frac{\varphi_j^\xi \varphi_j^{0\tau}}{r_j^0} \Phi^0(j, 0) \right\|$$

$$+\left(\sum_{j=N+1}^{n} \frac{\|\varphi_j^{\xi}\|^2}{r_j^0}\right)^{1/2}\left(\sum_{j=N+1}^{n} \frac{\|\Phi^{0^{\tau}}(j, 0)\varphi_j^0\|^2}{r_j^0}\right)^{1/2} \xrightarrow[\substack{n\to\infty\\ N\to\infty}]{} 0, \tag{56}$$

From (50) we know that there exists N such that

$$\left|\sqrt{\frac{r_j^0}{r_j}} - \sqrt{\frac{r_j}{r_j^0}}\right| \le 2, \quad \forall \quad j \ge N.$$

Hence for the last term on the right-hand side of (53) we can estimate as follows:

$$\left\|\sum_{j=0}^{n} \frac{\Phi(n+1, j+1)\varphi_j}{r_j^{1/2}}\left(\sqrt{\frac{r_j^0}{r_j}} - \sqrt{\frac{r_j}{r_j^0}}\right)\frac{\varphi_j^{0\tau}\Phi^0(j, 0)}{r_j^{0\,1/2}}\right\|$$

$$\le \left\|\sum_{j=0}^{N} \frac{\Phi(n+1, j+1)\varphi_j}{r_j^{1/2}}\left(\sqrt{\frac{r_j^0}{r_j}} - \sqrt{\frac{r_j}{r_j^0}}\right)\frac{\varphi_j^{0^{\tau}}\Phi^0(j, 0)}{r_j^{0\,1/2}}\right\|$$

$$+ 2\left(\sum_{j=N+1}^{n} \frac{\|\Phi(n+1, j+1)\varphi_j\|^2}{r_j}\right)^{1/2}$$

$$\times \left(\sum_{j=N+1}^{n} \frac{\|\varphi_j^{0^{\tau}}\Phi^0(j, 0)\|^2}{r_j^0}\right)^{1/2} \xrightarrow[\substack{n\to\infty\\ N\to\infty}]{} 0. \tag{57}$$

Combining (53)-(57) we conclude that $\Phi(n, 0) \to 0$ implies $\Phi^0(n, 0) \to 0$.

Conversely, assume that $\Phi^0(n, 0) \to 0$. By the argument similar to that used above, we are convinced of $r_n^0 \to \infty$. From here we can conclude $r_n \to \infty$; otherwise by (24) $\Sigma_{j=1}^{\infty} \|\varphi_j^{\xi}\|^2$ and $\Sigma_{j=1}^{\infty} \varphi_j^{\tau}\varphi_j^{\xi}$ would converge and r_n^0 would be bounded since

$$r_n^0 = r_n - 2\sum_{i=1}^{n} \varphi_j^{\tau}\varphi_j^{\xi} + \sum_{i=1}^{n} \|\varphi_j^{\xi}\|^2.$$

Then (50) still holds true by $r_n \to \infty$. Instead of (52) we now have

$$\Phi(n+1, 0) = \Phi^0(n+1, 0)$$

$$+ \sum_{j=0}^{n} \Phi^0(n+1, j+1)\left(\frac{\varphi_j^0 \varphi_j^{0\tau}}{r_j^0} - \frac{\varphi_j \varphi_j^\tau}{r_j}\right)\Phi(j, 0),$$

and by a completely similar argument we assert $\Phi(n, 0) \xrightarrow[n\to\infty]{} 0$. ∎

Remark 2.2. Theorem 2.3 actually has established the following more general fact: Let $\{\varphi_n^1\}$, $\{\varphi_n^2\}$, and $\{\psi_n\}$ be three arbitrary random vector sequences, if $\varphi_n^1 = \varphi_n^2 + \psi_n$ and

$$\sum_{n=0}^{\infty} \frac{\|\psi_n\|^2}{r_{1n}} < \infty,$$

then $\Phi_1(n, 0) \to 0$ if and only if $\Phi_2(n, 0) \to 0$, where $\Phi_i(n, 0)$ and r_{in} $(i = 1, 2)$ are defined by

$$\Phi_i(n+1, 0) = \left(I - \frac{\varphi_n^i \varphi_n^{i\tau}}{r_{in}}\right)\Phi_i(n, 0), \quad \Phi_i(0, 0) = I, \quad i = 1, 2, \tag{58}$$

$$r_{in} = 1 + \sum_{j=1}^{n} \|\varphi_j^i\|^2, \quad r_{i0} = 1, \quad i = 1, 2. \tag{59}$$

Let λ_{max}^{0n} $\left(\lambda_{min}^{0n}\right)$ denote the maximum (minimum) eigenvalue of

$$\sum_{j=1}^{n} \varphi_j^0 \varphi_j^{0\tau} + \frac{1}{d} I.$$

Theorem 2.4

If $C(z) - \frac{1}{2} I$ is strictly positive real and if

$$\lambda_{max}^n / \lambda_{min} \le M(\log r_n)^{1/4},$$

$$\forall\, n \ge N \quad \text{and} \quad r_n \to \infty, \quad \overline{\lim_{n\to\infty}}\, r_n / r_{n-1} < \infty$$

or

$$\lambda_{max}^{0n}/\lambda_{min}^{0n} \leq M\left(\log r_n^0\right)^{1/4},$$

$$\forall\, n \geq N \quad \text{and} \quad r_n^0 \to \infty, \quad \overline{\lim_{n\to\infty}}\, r_n^0/r_{n-1}^0 < \infty,$$

with N_0 and M possibly depending on ω, then

$$\theta_n \xrightarrow[n\to\infty]{} \theta \quad \text{a.s.}$$

for any initial value θ_0.

Proof. This theorem directly follows from Theorems 2.1, 2.2, 2.3.

III. EXISTENCE OF ADAPTIVE CONTROL AND ITS OPTIMALITY

In this section we consider the adaptive tracking problem; that is, the control u_n is given in order that the output y_n tracks a given deterministic reference sequence y_n^* when the system's parameters are unknown.

From (1), (5), and (21) we have

$$y_{n+1} = \theta^\tau \varphi_n^0 + w_{n+1};$$

hence

$$\hat{y}_{n+1} \triangleq E[y_{n+1} \mid \mathscr{F}_n] = \theta^\tau \varphi_n^0.$$

The simple calculation leads to

$$E\left(y_{n+1} - y_{n+1}^*\right)\left(y_{n+1} - y_{n+1}^*\right)^\tau$$

$$= E\left(y_{n+1} - \hat{y}_{n+1} + \hat{y}_{n+1} - y_{n+1}^*\right)\left(y_{n+1} - \hat{y}_{n+1} + \hat{y}_{n+1} - y_{n+1}^*\right)^\tau$$

$$= E\left(y_{n+1} - \hat{y}_{n+1}\right)\left(y_{n+1} - \hat{y}_{n+1}\right)^\tau$$

$$+ E\left(\hat{y}_{n+1} - y_{n+1}^*\right)\left(y_{n+1} - \hat{y}_{n+1}\right)^\tau$$

$$+ E\left(y_{n+1} - \hat{y}_{n+1}\right)\left(\hat{y}_{n+1} - y^*_{n+1}\right)^\tau$$

$$+ E\left(\hat{y}_{n+1} - y^*_{n+1}\right)\left(\hat{y}_{n+1} - y^*_{n+1}\right)^\tau$$

$$= Ew_{n+1}w^\tau_{n+1} + E\left(\hat{y}_{n+1} - y^*_{n+1}\right)\left(\hat{y}_{n+1} - y^*_{n+1}\right)^\tau$$

$$\geq Ew_{n+1}w^\tau_{n+1},$$

where the equality holds if and only if $\hat{y}_{n+1} = y^*_{n+1}$; that is, the control u_n is defined from $\theta^\tau\varphi^0_n = y^*_{n+1}$. If $Ew_{n+1}w^\tau_{n+1} = R$, then the minimum tracking error matrix also is R:

$$\min_{u_n \in \mathscr{F}_n} E\left(y_{n+1} - y^*_{n+1}\right)\left(y_{n+1} - y^*_{n+1}\right)^\tau = R.$$

When θ and φ^0_n are unavailable, then they are naturally replaced by θ_n and φ_n, respectively. Hence in this case the adaptive tracking control u_n should be defined from

$$\theta^\tau_n\varphi_n = y^*_{n+1}. \tag{60}$$

We first consider the existence problem for u_n satisfying (60) and then show its optimality. Actually we proceed to discuss for u_n an equation more general than (60), which will be needed in the later sections, namely,

$$\theta^\tau_n\varphi_n = y^*_{n+1} + v_n, \tag{61}$$

where v_n is an arbitrary $\mathscr{F}_n$-measurable and m-dimensional disturbance sequence. When $v_n \equiv 0$, (61) concides with (60).

Theorem 3.1

Assume that $m \leq l$ and $\{w_n\}$ and $\{v_n\}$ are two mutually independent sequences of random vectors and that the components of w_n are independent and have continuous-type distributions. Then for any $n \geq 1$, there exists u_n satisfying (61) if the initial values are appropriately chosen. Further, this u_n is unique if and only if $m = l$.

We first prove a lemma.

Lemma 3.1. (1) Let A and B be two matrices of dimensions $m \times n$ and $n \times m$, respectively. Then the following equality takes place:

$$\det(I_m + AB) = \det(I_n + BA),$$

where I_n means the $n \times n$ identity matrix.

(2) Provided that x_1 and x_2 are independent random variables,

$$\sup_{a \in R^1} P(x_1 + x_2 = a) \le \min\{\sup_{a \in R^1} P(x_1 = a), \sup_{a \in R^1} P(x_2 = a)\}.$$

Proof. (1) By taking determinants for both sides of the matrix identity

$$\begin{bmatrix} I_m & -A \\ 0 & BA + I_n \end{bmatrix} = \begin{bmatrix} I_m & 0 \\ -B & I_n \end{bmatrix}\begin{bmatrix} I_m + AB & -A \\ 0 & I_n \end{bmatrix}\begin{bmatrix} I_m & 0 \\ B & I_n \end{bmatrix},$$

the desired equality is immediately verified.

(2) Denote by $F_1(x)$, $F_2(x)$, and $F_{12}(x)$ the distributions of x_1, x_2, and $x_1 + x_2$, respectively. Clearly, we have

$$F_{12}(x) = \int_{-\infty}^{\infty} F_1(x - y)\, dF_2(y)$$

and

$$F_{12}(x+) = \int_{-\infty}^{\infty} F_1((x - y)+)\, dF_2(y)$$

by the dominated convergence theorem.

Then for any $a \in R^1$

$$P(x_1 + x_2 = a) = F_{12}(a+) - F_{12}(a)$$

$$= \int_{-\infty}^{\infty} [F_1((a - y)+) - F_1(a - y)]\, dF_2(y)$$

$$= \int_{-\infty}^{\infty} P(x_1 = a - y)\, dF_2(y) \leq \sup_{a \in R^1} P(x_1 = a)$$

$$\times \int_{-\infty}^{\infty} dF_2(y) = \sup_{a \in R^1} P(x_1 = a).$$

Similarly, we have $P(x_1 + x_2 = a) \leq \sup_{a \in R^1} P(x_2 = a)$ and the desired result follows. ■

Proof of Theorem 3.1. Let A_{in}, B_{jn}, C_{kn}, $i = 1, \ldots, p$, $j = 1, \ldots, q$, $k = 1, \ldots, r$ be the matrix components of θ_n, that is,

$$\theta_n^\tau \triangleq [-A_{1n} \cdots -A_{pn} \quad B_{1n} \cdots B_{qn} \quad C_{1n} \cdots C_{rn}].$$

Set

$$\overline{\theta}_n^\tau \triangleq [-A_{1n} \cdots -A_{pn} \quad 0 \quad B_{2n} \cdots B_{qn} \quad C_{1n} \cdots C_{rn}]$$

and

$$\overline{\varphi}_n \triangleq \Big[y_n^\tau \cdots y_{n-p+1}^\tau \quad 0 \quad u_{n-1}^\tau \cdots u_{n-q+1}^\tau$$

$$y_n^\tau - \varphi_{n-1}^\tau \theta_{n-1} \quad \cdots \quad y_{n-r+1}^\tau - \varphi_{n-r}^\tau \theta_{n-r}\Big]^\tau.$$

Equation (61) is equivalent to

$$B_{1n} u_n = y_{n+1}^* + v_n - \overline{\theta}_n^\tau \overline{\varphi}_n. \tag{62}$$

First let $m = l$. For this case we only need to prove that B_{1n} is invertible a.s. In fact, if this is true, then from (62) u_n is uniquely defined by $u_n = B_{1n}^{-1}\left(y_{n+1}^* + v_n - \overline{\theta}_n^\tau \overline{\varphi}_n\right)$, which obviously is $\mathscr{F}_n$ measurable.

From (6) and (13) we obtain

$$B_{1n+1} = B_{1n} + \frac{1}{r_n}(\xi_{n+1} + w_{n+1}) u_n^\tau. \tag{63}$$

It is easy to take initial values φ_0, θ_0 such that B_{11} is invertible; for example, take $u_0 = 0$ and B_{10} invertible.

We now inductively prove that B_{1n} is nondegenerate for any $n \geq 0$. Assume that B_{1n} is nonsingular a.s. We show that so is B_{1n+1}; in other words, we need to prove that $P(N) = 0$ implies $P(DN^c) = 0$, where

$$N \triangleq \{\omega \mid \det B_{1n} = 0\}, \quad D \triangleq \{\omega \mid \det B_{1n+1} = 0\}.$$

Suppose that the opposite were true, that is, $P(N) = 0$, but $P(DN^c) > 0$.

From (63) we have

$$\det\left(B_{1n} + \frac{1}{r_n}(\xi_{n+1} + w_{n+1})u_n^\tau\right) = 0, \quad \forall\ \omega \in DN^c,$$

but for $\omega \in DN^c$, $\det B_{1n} \neq 0$. Hence

$$\det\left(I + \frac{1}{r_n} B_{1n}^{-1}(\xi_{n+1} + w_{n+1})u_n^\tau\right) = 0, \quad \forall\ \omega \in DN^c,$$

or

$$\det\left(1 + \frac{1}{r_n} u_n^\tau B_{1n}^{-1}(\xi_{n+1} + w_{n+1})\right) = 0, \quad \forall\ \omega \in DN^c,$$

by part (1) of Lemma 3.1.

Then we have

$$u_n B_{1n}^{-1}(\xi_{n+1} + w_{n+1}) = -r_n, \quad \forall\ \omega \in DN^c. \tag{64}$$

Consequently,

$$u_n^\tau B_{1n}^{-1} \neq 0, \quad \forall\ \omega \in DN^c \tag{65}$$

since $r_n \geq 1$.

We denote by $\alpha_i(\omega)$ and $w_{n+1,i}$ the components of $u_n^\tau B_{1n}^{-1}$ and w_{n+1}, respectively; that is,

$$u_n^\tau B_{1n}^{-1} = [\alpha_1(\omega), \ldots, \alpha_m(\omega)], \tag{66}$$

$$w_{n+1} = [w_{n+1,1}, \ldots, w_{n+1,m}]^\tau. \tag{67}$$

Then from (64), (66), and (67) we have

$$\sum_{i=1}^{m} \alpha_i(\omega) w_{n+1,i} + r_n + u_n^\tau B_{1n}^{-1}\xi_{n+1} = 0, \quad \forall\ \omega \in DN^c. \tag{68}$$

From (65) and the assumption $P(DN^c) > 0$ we would have some $\alpha_i(\omega)$ and a subset $D_1 \subset DN^c$ such that

$$\alpha_i(\omega) \neq 0 \quad \forall\, \omega \in D_1, \quad P(D_1) > 0. \tag{69}$$

Without loss of generality, we assume i = 1, and define the random variable z

$$z(\omega) = \begin{cases} \dfrac{1}{\alpha_1(\omega)}\left[\sum_{i=2}^{m} \alpha_i(\omega) w_{n+1,i} + r_n + u_n^\tau B_{1n}^{-1}\xi_{n+1}\right], & \omega \in D_1, \\ 0, & \omega \in D_1^c, \end{cases}$$

which is clearly indpendent of $w_{n+1,1}$. By part (2) of Lemma 3.1 it follows that

$$P(w_{n+1,1} + z(\omega) = 0) = 0. \tag{70}$$

However, (68) and (69) would yield

$$P(w_{n+1,1} + z(\omega) = 0) \geq P(D_1) > 0. \tag{71}$$

The contradiction obtained proves $P(DN^c) = 0$, and hence the nonsingularity of B_{1n+1} a.s.

Now assume $m < l$. Let

$$B_{1n} \triangleq \Big[\underbrace{B_{1n}^1}_{m}, \underbrace{B_{1n}^2}_{l-m}\Big], \qquad u_n^\tau = \Big[\underbrace{u_n^{1\tau}}_{m}, \underbrace{u_n^{2\tau}}_{l-m}\Big].$$

From (63) we see

$$B_{1n+1}^1 = B_{1n}^1 + \frac{1}{r_n}(\xi_{n+1} + w_{n+1})u_n^{1\tau}.$$

In a similar way as given for the $l = m$ case we can prove that B_{1n}^1 is invertible a.s. for any $n \geq 1$ if φ_0 and θ_0 are adequately chosen. Then (62) is equivalent to

$$\left[I, \left(B_{1n}^1\right)^{-1}B_{1n}^2\right]u_n = \left(B_{1n}^1\right)^{-1}\left(y_{n+1}^* + v_n - \bar{\theta}_n^\tau\bar{\varphi}_n\right)$$

or (72)

$$u_n^1 + \left(B_{1n}^1\right)^{-1}B_{1n}^2 u_n^2 = \left(B_{1n}^1\right)^{-1}\left(y_{n+1}^* + v_n - \bar{\theta}_n^\tau\bar{\varphi}_n\right).$$

Then, obviously, the solution of (72) can be expressed by

$$u_n = \begin{bmatrix} \left(B_{1n}^1\right)^{-1}\left(y_{n+1}^* + v_n - \overline{\theta}_n^\tau \overline{\varphi}_n - B_{1n}^2 u_n^2\right) \\ u_n^2 \end{bmatrix}, \quad \text{a.s.}$$

with any $(l - m)$-dimensional and $\mathscr{F}_n$-measurable u_n^2. This means for the case $m < l$ the control u_n satisfying (61) exists but it is not unique. ■

Remark 3.1. Recently Caines and Meyn [24] also have shown the existence of u_n satisfying (60) for the one-dimensional case but under conditions different from those imposed here.

In the sequel we always assume that adaptive control exists. For the stability of adaptive control systems and the optimality of tracking, we need the following conditions:

(A_1) $C(z) - \frac{1}{2} I$ is strictly positive real.

(A_2) B_1 is of full rank and the zeros of det $B_1^+ B(z)$ lie outside the closed unit disk and $p \geq 1$, $q \geq 1$, and $m \geq l$, where p, q, m, and l are defined in (1).

(A_3) u_n is selected to satisfy (60), where $\{y_n^*\}$ is a bounded deterministic reference sequence.

Theorem 3.2

For system (1) and the algorithm defined by (6) through (8), suppose that conditions (A_1)-(A_3) hold and

$$\lim_{n\to\infty} \frac{1}{n} \sum_{i=1}^{n} w_i w_i^\tau = R \neq 0. \tag{73}$$

Then $r_n \to \infty$ and the system is with properties

$$\overline{\lim_{n\to\infty}} \frac{1}{n} \sum_{i=1}^{n} \|u_i\|^2 < \infty, \quad \text{a.s.} \tag{74}$$

$$\overline{\lim_{n\to\infty}} \frac{1}{n} \sum_{i=1}^{n} \|y_i\|^2 < \infty, \quad \text{a.s.} \tag{75}$$

$$\lim_{n\to\infty} \frac{1}{n} \sum_{i=1}^{n} \left(y_i - y_i^*\right)\left(y_i - y_i^*\right)^\tau = R, \quad \text{a.s.} \tag{76}$$

Proof. We first note that by (8) $\lim_{n\to\infty} r_n < \infty$ implies $\varphi_n \to 0$ and hence $y_n \to 0$ and $u_n \to 0$, which in turn imply $w_n \to 0$, since $C(z)$ is asymptotically stable as a consequence of the strictly positive realness assumption for $C(z) - \frac{1}{2} I$. Clearly, $w_n \to 0$ yields

$$\frac{1}{n} \sum_{i=1}^{n} w_i w_i^\tau \to 0,$$

which contradicts with (73). Thus we conclude $r_n \to \infty$.

By Lemma 2.1 and the Kronecker lemma we have

$$\frac{1}{r_n} \sum_{i=1}^{n} \|\xi_{i+1}\|^2 \xrightarrow[n\to\infty]{} 0, \quad \text{a.s.} \tag{77}$$

Then by (1), (73), and condition (A_2) it follows that

$$\frac{1}{n} \sum_{i=1}^{n} \|u_i\|^2 \le \frac{1}{n} \sum_{i=1}^{n} \|y_{i+1}\|^2 + k_2, \tag{78}$$

where and hereafter k_i always denotes a positive quantity that is constant in time and possibly depending on ω.

By (13) and (73) we have

$$\frac{1}{n} \sum_{i=1}^{n} \|y_{i+1} - \theta_i^\tau \varphi_i\|^2 \le \frac{k_3}{n} \sum_{i=1}^{n} \|\xi_{i+1}\|^2 + k_4.$$

Hence by condition (A_3) we see that

$$\frac{1}{n} \sum_{i=1}^{n} \|y_{i+1}\|^2 \le \frac{k_5}{n} \sum_{i=1}^{n} \|\xi_{i+1}\|^2 + k_6. \tag{79}$$

Thus, we have the estimate

$$\frac{r_n}{n} \le \frac{k_7}{n} \sum_{i=1}^{n} \|\xi_{i+1}\|^2 + k_8 = \frac{k_7 r_n}{n}\left(\frac{1}{r_n} \sum_{i=1}^{n} \|\xi_{i+1}\|^2\right) + k_8,$$

which leads to

$$\frac{r_n}{n} \le \left(k_8 \Big/ \left(1 - k_7 \frac{1}{r_n} \sum_{i=1}^{n} \|\xi_{i+1}\|^2\right)\right) \xrightarrow[n\to\infty]{} k_8.$$

Thus we conclude that

$$\overline{\lim_{n\to\infty}} \frac{r_n}{n} < \infty \tag{80}$$

and by (77) that

$$\frac{1}{n} \sum_{i=1}^{n} \|\xi_{i+1}\|^2 \xrightarrow[n\to\infty]{} 0, \quad \text{a.s.} \tag{81}$$

Then (75) follows from (79) and (81), while (74) from (75) and (78).

By (13) and condition (A_3) it is easy to see that

$$\left(y_i - y_i^*\right)\left(y_i - y_i^*\right)^\tau = w_i w_i^\tau + \xi_i \xi_i^\tau + \xi_i w_i^\tau + w_i \xi_i^\tau.$$

From here by using (73) and (81) we obtain (76). ∎

IV. OPTIMAL ADAPTIVE CONTROL WITH CONSISTENT PARAMETER ESTIMATES

In the preceding section we have shown that adaptive control satisfying (60) is optimal, but, generally speaking, under this control the parameter estimates may be inconsistent. We explain this by taking $y_n^* \equiv 0$; in this case, (60) becomes $\theta_n^\tau \varphi_n \equiv 0$, which together with (6) leads to

$$\theta_n^\tau(\theta_{n+1} - \theta_n) = \theta_n^\tau \frac{\varphi_n}{r_n} y_{n+1}^\tau = 0,$$

and then

$$\theta_n^\tau\theta_n = [\theta_{n-1} + (\theta_n - \theta_{n-1})]^\tau[\theta_{n-1} + (\theta_n - \theta_{n-1})]$$

$$= \theta_{n-1}^\tau\theta_{n-1} + (\theta_n - \theta_{n-1})^\tau(\theta_n - \theta_{n-1})$$

$$= \theta_0^\tau\theta_0 + \sum_{i=1}^{n}(\theta_i - \theta_{i-1})^\tau(\theta_i - \theta_{i-1}) \geq \theta_0^\tau\theta_0.$$

Therefore, θ_n cannot be consistent if the initial value θ_0 is taken so that $\theta_0^\tau\theta_0 > \theta^\tau\theta$.

Now, instead of (60) we consider (61) where a disturbance $\{v_n\}$ is artificially introduced to the reference signal $\left\{y_n^*\right\}$. The main purpose of this section is to give the optimal adaptive control by which the long-run average of tracking error is minimal and the parameter estimates are strongly consistent. The idea is that we take the adaptive control u_n to satisfy (61) instead of (60) and take the disturbance sequence v_n tending to zero at such a rate that Theorem 2.4 can be applied.

We need the following conditions.

(A_4) $\{w_i\}$ and $\{v_i\}$ are two m-dimensional random sequences for which w_i, w_j, v_k, v_l are mutually independent for any $i \neq j$, $k \neq l$, with the properties $Ew_i = Ev_i = 0$, $\forall\ i \geq 0$; $w_i = v_i = 0$, $\forall\ i < 0$; and $\sup_i E\|w_i\|^{4+\delta} < \infty$, $\sup_i E\|v_i\|^{4+\delta} < \infty$ for some $\delta > 0$ and

$$\lim_{n\to\infty} \frac{1}{n} \sum_{i=1}^{n} w_i w_i^\tau = R > 0. \quad \text{a.s.} \tag{82}$$

(A_5) u_n is selected to satisfy (61) and y_m^* is a bounded deterministic reference sequence.

Lemma 4.1. Theorem 3.2 is still valid with (A_3) replaced by (A_5) if $\{v_n\}$ satisfies

$$\lim_{n\to\infty} \frac{1}{n} \sum_{i=1}^{n} \|v_i\|^2 = 0. \tag{83}$$

Proof. The proof is similar to that of Theorem 3.2.

Lemma 4.2. Let $\{v_n\}$ satisfy conditions in (A_4) and let

$$H_N(z) = \sum_{i=0}^{\infty} H_i(N) z^i$$

be the matrix series in shift-back operator z with

$$\|H_i(N)\| \le k_1 \exp(-k_2 i),$$

for $\forall\, i \ge 0$, $\forall\, N \ge 0$ and some $k_1 > 0$ and $k_2 > 0$. Then

$$\lim_{n\to\infty} \frac{\log^{1/4} N}{N} \sum_{n=1}^{N} (H_N(z)v_n)(H_N(z)v_n)^\tau$$

$$= \lim_{n\to\infty} \frac{\log^{1/4} N}{N} \sum_{i=0}^{N} H_i(N)\left(\sum_{n=1}^{N} R_n\right) H_i^\tau(N),$$

provided the limit on the left-hand side exists, where $R_n \triangleq E v_n v_n^\tau$.

Proof. Due to the assumption that $v_n = 0$ for $n < 0$, we have

$$\sum_{n=1}^{N} (H_N(z)v_n)(H_N(z)v_n)^\tau = \sum_{i,j=0}^{\infty} H_i(N)\left(\sum_{n=1}^{N} v_{n-i}v_{n-j}^\tau\right) H_j^\tau(N)$$

$$= \sum_{i,j=0}^{\infty} H_i(N)$$

$$\times \left(\sum_{n=\max(i,j,1)}^{N} v_{n-1}v_{n-j}^\tau\right) H_j^\tau(N)$$

Set

$$S_N(i, j) = \sum_{n=\max(i,j,1)}^{N} \left[v_{n-i}v_{n-j}^{\tau} - \delta_{ij}R_{n-i}\right],$$

$$\delta_{ij} = \begin{cases} 1, & i = j \\ 0, & i \neq j. \end{cases}$$

Clearly, $S_N(i, j)$ is a martingale and by Burkholder inequality [25], C_r inequality, and Schwarz inequality, we have for any $i \geq 0$, $j \geq 0$,

$$\begin{aligned} E\|S_N(i, j)\|^{2+\delta/2} &\leq c_1 E\left(\sum_{n=\max(i,j,1)}^{N} \|v_{n-i}v_{n-j}^{\tau} - \delta_{ij}R_{n-i}\|^2\right)^{1+\delta/4} \\ &\leq c_1 N^{\delta/4} E \sum_{n=\max(i,j,1)}^{N} \|v_{n-i}v_{n-j}^{\tau} - \delta_{ij}R_{n-i}\|^{2+\delta/2} \\ &\leq c_2 N^{1+\delta/4}, \end{aligned}$$

for some $c_1 > 0$, $c_2 > 0$. From this and the Hölder inequality, it follows that for any $\epsilon > 0$ and

$$\gamma \in \left(\frac{2 + \delta/4}{2 + \delta/2}, 1\right)$$

that

$$P\left\{\left\|\sum_{i,j=0}^{\infty} H_i(N)S_N(i, j)H_j^{\tau}(N)\right\| > N^{\gamma} \cdot \epsilon\right\}$$

$$\leq \frac{1}{\epsilon^{2+\delta/2} \cdot N^{\gamma(2+\delta/2)}} E\left(\sum_{i,j=0}^{\infty} \|H_i(N)\| \cdot \|H_j(N)\| \cdot \|S_N(i, j)\|\right)^{2+\delta/2}$$

$$\leq c_3 \frac{1}{N^{\gamma(2+\delta/2)}} E \sum_{i,j=0}^{\infty} (\|H_i(N)\| \cdot \|H_j(N)\|)^{1+\delta/4}$$

$$\cdot \|S_N(i, j)\|^{2+\delta/2}$$

$$\leq c_4 \frac{1}{N^{\gamma(2+\delta/2)-(1+\delta/4)}},$$

for any $N \geq 1$ and some constants $c_3 > 0$ and $c_4 > 0$.

Then Borel-Cantelli lemma gives

$$\lim_{N\to\infty} \frac{1}{N^{\gamma}} \sum_{i,j=0}^{\infty} H_i(N) S_N(i, j) H_j^{\tau}(N) = 0. \tag{84}$$

Finally, by (84) we obtain the desired result:

$$\lim_{N\to\infty} \frac{\log^{1/4} N}{N} \sum_{n=1}^{N} (H_N(z) v_n)(H_N(z) v_n)^{\tau}$$

$$= \lim_{N\to\infty} \frac{\log^{1/4} N}{N} \sum_{i,j=0}^{\infty} H_i(N) S_N(i, j) H_j^{\tau}(N)$$

$$+ \lim_{N\to\infty} \frac{\log^{1/4} N}{N} \sum_{i,j=0}^{\infty} H_i(N) \sum_{n=\max(i,j,1)}^{N} \delta_{ij} R_{n-i} H_j^{\tau}(N)$$

$$= \lim_{N\to\infty} \frac{\log^{1/4} N}{N} \sum_{i=0}^{N} H_i(N) \sum_{n=\max(i,1)}^{N} R_{n-i} H_i^{\tau}(N)$$

$$= \lim_{N\to\infty} \frac{\log^{1/4} N}{N} \left[H_0(N) \sum_{n=1}^{N} R_n H_0^{\tau}(N) + \sum_{i=1}^{N} H_i(N) \sum_{n=0}^{N-i} R_n H_i^{\tau}(N) \right]$$

$$= \lim_{N\to\infty} \frac{\log^{1/4} N}{N} \left[H_0(N) \sum_{n=1}^{N} R_n H_0^{\tau}(N) \right.$$

$$\left. + \sum_{i=1}^{N} H_i(N) \left(\sum_{n=1}^{N} R_n + R_0 - \sum_{n=N-i+1}^{N} R_n \right) H_i^{\tau}(N) \right]$$

$$= \lim_{N\to\infty} \frac{\log^{1/4} N}{N} \sum_{i=0}^{N} H_i(N) \sum_{n=1}^{N} R_n H_i^\tau(N). \quad \blacksquare$$

Lemma 4.3. Let all assumptions in (A_4) be satisfied except (82) and let $H(z) = \Sigma_{i=0}^{\infty} H_i z^i$ and $G(z) = \Sigma_{i=0}^{\infty} G_i z^i$ be matrix series in shift-back operator z with

$$\|H_i\| + \|G_i\| \le k_1 \exp(-k_2 i),$$

for any $i \ge 0$ and some constants $k_1 > 0$ and $k_2 > 0$. Then there exists $\gamma \in (0, 1)$ such that for $\forall\, l \ge 0$, $m \ge 0$,

$$\lim_{N\to\infty} \frac{1}{N^\gamma} \sum_{n=1}^{N} (H(z)w_{n+1-l})(G(z)v_{n-m})^\tau = 0, \quad \text{a.s.}, \tag{85}$$

$$\overline{\lim_{N\to\infty}} \frac{1}{N} \sum_{n=1}^{N} \|H(z)w_{n+1-l}\|^2 < \infty, \quad \text{a.s.}, \tag{86}$$

and

$$\lim_{N\to\infty} \frac{1}{N^\gamma} \sum_{n=1}^{N} (H(z)w_{n+1-l})\eta_n^\tau = 0, \quad \text{a.s.}, \tag{87}$$

for any bounded deterministic sequence $\{\eta_n\}$.

Proof. Set

$$S_N(i, j) \triangleq \sum_{n=1}^{N} w_{n+1-l-i} v_{n-m-j}^\tau.$$

Similar to the proof of (84), one can easily be persuaded of

$$\lim_{N\to\infty} \frac{1}{N^\gamma} \sum_{i,j=0}^{\infty} H_i S_N(i, j) G_j^\tau = 0,$$

from which (85) follows immediately.

Clearly, (87) can be verified in a similar fashion.

By setting $H_N(z) \equiv H(z)$ and $v_n = w_{n+1-l}$ in (84), we have

$$\lim_{N\to\infty} \frac{1}{N^\gamma} \sum_{i,j=0}^{\infty} H_i S_N(i, j) H_j^\tau = 0,$$

where

$$S_N(i, j) = \sum_{n=\max(i,j,1)}^{N} \left[w_{n-l+1-i} w_{n-l+1-j}^\tau - \delta_{ij} R_{n-l+1-i}\right]$$

and

$$R_{n-l+1-i} = E w_{n-l+1-i} w_{n-l+1-i}^\tau.$$

Then by the uniform boundness of R_n, we have

$$\overline{\lim_{N\to\infty}} \frac{1}{N} \sum_{n=1}^{N} \|H(z) w_{n+1-l}\|^2$$

$$= \overline{\lim_{N\to\infty}} \operatorname{tr}\left[\frac{1}{N} \sum_{n=1}^{N} (H(z) w_{n+1-l})(H(z) w_{n+1-l})^\tau\right]$$

$$= \overline{\lim_{N\to\infty}} \operatorname{tr}\left[\frac{1}{N} \sum_{n=1}^{N} H_i S_N(i, j) H_j^\tau \right.$$

$$\left. + \sum_{i,j=0}^{\infty} H_i \frac{1}{N} \sum_{n=\max(i,j,1)}^{N} \delta_{ij} R_{n-l+1-i} H_j^\tau\right]$$

$$= \overline{\lim_{N\to\infty}} \operatorname{tr} \sum_{i=0}^{\infty} H_i \frac{1}{N} \sum_{n=\max(i,1)}^{N} R_{n-l+1-i} H_i^\tau < \infty.$$

This completes the proof of the lemma. ■

Set

$$H_1(z) = \left[B_1^+ B(z)\right]^{-1} B_1^+ A(z), \tag{88}$$

$$H_2(z) = H_1(z) - \left[B_1^+ B(z)\right]^{-1} B_1^+ C(z), \tag{89}$$

$$Y_n^* = \left[y_n^{*\tau} \cdots y_{n-p+1}^{\tau} \quad \left(H_1(z) y_{n+1}^*\right)^{\tau} \cdots \left(H_1(z) y_{n-q+2}^*\right)^{\tau} \right]^{\tau}, \tag{90}$$

and

$$z_n = \left[v_{n-1}^{\tau} \cdots v_{n-p}^{\tau} \quad (H_1(z) v_n)^{\tau} \cdots (H_1(z) v_{n-q+1})^{\tau} \right]^{\tau}. \tag{91}$$

In the following by $\lambda_{min}(X)$ ($\lambda_{max}(X)$) we mean the minimum (maximum) eigenvalue of the matrix X.

Lemma 4.4. For system (1) and the algorithm defined by (5) through (8), if conditions (A_1), (A_2), (A_4), and (A_5) are fulfilled and if

$$\underline{\lim}_{N\to\infty} \lambda_{min}\left(\frac{\log^{1/4} N}{N} \sum_{n=1}^{N} \left(Y_n^* Y_n^{*\tau} + z_n z_n^{\tau} \right) \right) \neq 0, \quad \text{a.s.,} \tag{92}$$

then the parameter estimate θ_n is strongly consistent:

$$\theta_n \xrightarrow[n\to\infty]{} \theta, \quad \text{a.s.}$$

Proof. To begin with, we explain the meaning of the lemma. Clearly, we have

$$\overline{\lim}_{N\to\infty} \lambda_{max}\left(\frac{1}{N} \sum_{n=1}^{N} z_n z_n^{\tau} \right) < \infty$$

[see (86)], we assume it to be greater than 0. If

$$\underline{\lim}_{N\to\infty} \lambda_{min}\left(\frac{1}{N} \sum_{n=1}^{N} z_n z_n^{\tau} \right) = 0,$$

then the persistent excitation condition does not hold. The lemma implies that θ_n still tends to θ if

$$\lambda_{min}\left(\frac{1}{N} \sum_{n=1}^{N} z_n z_n^{\tau} \right)$$

goes to 0 at a rate not faster than $0(\log^{-1/4} N)$.

By Theorems 2.1 and 2.3 we only need to prove $\Phi^0(n, 0) \xrightarrow[n\to\infty]{} 0$ a.s., where $\Phi^0(n, 0)$ is defined by (47). By (1) we have

$$u_n = \left[B_1^+ B(z)\right]^{-1} B_1^+ A(z) y_{n+1} - \left[B_1^+ B(z)\right]^{-1} B_1^+ C(z) w_{n+1}. \tag{93}$$

Then by (13), (61), (88), (89), and (93), φ_n^0 defined by (21) can be written as

$$\varphi_n^0 = \varphi_n^1 + \psi_n, \quad \varphi_n^1 = \varphi_n^2 + \varphi_n^3, \tag{94}$$

where

$$\psi_n = \left[\xi_n^\tau \cdots \xi_{n-p+1}^\tau (H_1(z)\xi_{n+1})^\tau \cdots (H_1(z)\xi_{n-q+2})^\tau 0 \cdots 0\right]^\tau, \tag{95}$$

$$\varphi_n^2 = \left[w_n^\tau \cdots w_{n-p+1}^\tau, (H_2(z)w_{n+1})^\tau \cdots (H_2(z)w_{n-q+2})^\tau, w_n^\tau \cdots w_{n-r+1}^\tau\right]^\tau, \tag{96}$$

$$\varphi_n^3 = \left[Y_n^{*\tau} + z_n^\tau, 0 \cdots 0\right]^\tau. \tag{97}$$

By Lemma 4.1 we know that $r_n \to \infty$, and hence (51) is still true. Then by the Schwarz inequality it follows that

$$\sum_{n=0}^{\infty} \frac{\|H_1(z)\xi_{n+1-l}\|^2}{r_n^0} \le \sum_{n=0}^{\infty} \frac{1}{r_n^0} \sum_{i=0}^{\infty} \|H_{1i}\| \sum_{i=0}^{\infty} \|H_{1i}\| \cdot \|\xi_{n+1-l-i}\|^2$$
$$\le k_0 \sum_{i=0}^{\infty} \|H_{1i}\| \sum_{n=0}^{\infty} \frac{\|\xi_{n+1-l-i}\|^2}{r_n^0} < \infty,$$

where the last inequality is obtained because $\xi_i = 0$ for $i < 0$ and the coefficients in

$$H_1(z) \sum_{i=0}^{\infty} H_{1i} z^i$$

have the estimates

$$\|H_{1i}\| \le k_1 \exp(-k_2 i), \quad \forall \ i \ge 0,$$

and some $k_1 > 0$, $k_2 > 0$ by condition (A_2). Thus we have established that

$$\sum_{n=0}^{\infty} \frac{\|\psi_n\|^2}{r_n^0} < \infty. \tag{98}$$

Then from (94), (98), and Remark 2.2, we conclude that $\Phi^0(n, 0) \xrightarrow[n\to\infty]{} 0$ if and only if $\Phi_1(n, 0) \xrightarrow[n\to\infty]{} 0$.

Next, we prove that

$$\underline{\lim}_{N\to\infty} \lambda_{\min}\left(\frac{\log^{1/4} N}{N} \sum_{n=1}^{N} \varphi_n^1 \varphi_n^{1\tau}\right) \neq 0. \tag{99}$$

If (99) were not true, then we would find a subsequence of eigenvectors

$$\begin{bmatrix} \alpha_{N_k} \\ \beta_{N_k} \end{bmatrix}$$

for matrices

$$\frac{\log^{1/4} N_k}{N_k} \sum_{n=1}^{N_k} \varphi_n^1 \varphi_n^{1\tau}$$

with $N_k \xrightarrow[k\to\infty]{} \infty$, $\alpha_{N_k} \in R^{mp+lq}$, $\beta_{N_k} \in R^{mr}$ and

$$\|\alpha_{N_k}\|^2 + \|\beta_{N_k}\|^2 = 1 \tag{100}$$

such that

$$\left(\alpha_{N_k}^\tau, \beta_{N_k}^\tau\right) \frac{\log^{1/4} N_k}{N_k} \sum_{n=1}^{N_k} \varphi_n^1 \varphi_n^{1\tau} \begin{pmatrix} \alpha_{N_k} \\ \beta_{N_k} \end{pmatrix} \xrightarrow[k\to\infty]{} 0. \tag{101}$$

Utilizing Lemma 4.3, one can easily be persuaded of the fact that

$$\frac{\log^{1/4} N}{N} \sum_{n=1}^{N} \varphi_n^2 \varphi_n^{3\tau} \xrightarrow[N\to\infty]{} 0. \tag{102}$$

Then (101) is reduced to

$$\left(\alpha_{N_k}^{\tau},\ \beta_{N_k}^{\tau}\right)\frac{\log^{1/4} N_k}{N_k}\sum_{n=1}^{N_k}\varphi_n^2\varphi_n^{2\tau}\begin{pmatrix}\alpha_{N_k}\\ \beta_{N_k}\end{pmatrix}\xrightarrow[k\to\infty]{} 0. \tag{103}$$

$$\left(\alpha_{N_k}^{\tau},\ \beta_{N_k}^{\tau}\right)\frac{\log^{1/4} N_k}{N_k}\sum_{n=1}^{N_k}\varphi_n^3\varphi_n^{3\tau}\begin{pmatrix}\alpha_{N_k}\\ \beta_{N_k}\end{pmatrix}\xrightarrow[k\to\infty]{} 0. \tag{104}$$

In view of Lemma 4.3, (92) implies that

$$\underline{\lim}_{N\to\infty}\lambda_{\min}\left(\frac{\log^{1/4} N}{N}\sum_{n=1}^{N}\left(Y_n^* + z_n\right)\left(Y_n^* + z_n\right)^{\tau}\right)\neq 0. \tag{105}$$

Noting the fact that the last mr elements in φ_n^3 are zeros, by (104) and (105), we conclude that

$$\alpha_{N_k}\xrightarrow[k\to\infty]{} 0. \tag{106}$$

Whence, recalling (100), we have

$$\|\beta_{N_k}\|\xrightarrow[k\to\infty]{} 1. \tag{107}$$

Let

$$x_n^1 = \left[w_n^{\tau}\ \cdots\ w_{n-p+1}^{\tau}\quad (H_2(z)w_{n+1})^{\tau}\ \cdots\ (H_2(z)w_{n-q+2})^{\tau}\right]^{\tau},$$

$$x_n^2 = \left[w_n^{\tau}\ \cdots\ w_{n-r+1}^{\tau}\right]^{\tau}.$$

Then $\varphi_n^2 = \left[x_n^{1\tau}\quad x_n^{2\tau}\right]^{\tau}$ and (103) implies that

$$\frac{1}{N_k}\sum_{n=1}^{N_k}\|\alpha_{N_k}^{\tau}x_n^1 + \beta_{N_k}^{\tau}x_n^2\|^2\xrightarrow[k\to\infty]{} 0. \tag{108}$$

Further, we have

$$\overline{\lim}_{N\to\infty}\frac{1}{N}\sum_{n=1}^{N}\|x_n^1\|^2 < \infty \tag{109}$$

by Lemma 4.3, and

$$\lim_{N\to\infty} \frac{1}{N} \sum_{n=1}^{N} x_n^2 x_n^{2\tau} = \begin{bmatrix} R & & 0 \\ & \ddots & \\ 0 & & R \end{bmatrix} > 0 \tag{110}$$

by ergodicity.

Thus from (106) and (108) through (110), it follows that

$$\lim_{k\to\infty} \frac{1}{N_k} \sum_{n=1}^{N_k} \beta_{N_k}^{\tau} x_n^2 x_n^{2\tau} \beta_{N_k} = 0,$$

which leads to $\beta_{N_k} \xrightarrow[k\to\infty]{} 0$ by (110). Comparing it with (107) we obtain a contradiction that shows the truth of (99). Therefore, there exist $\alpha_0 > 0$ and N_0 such that

$$\lambda_{\min}\left(\sum_{i=1}^{n} \varphi_i^1 \varphi_i^{1\tau}\right) \geq \frac{n}{\log^{1/4} n} \alpha_0, \quad \forall \quad n \geq N_0. \tag{111}$$

By (94) and Lemma 4.3 it follows that

$$\varliminf_{n\to\infty} \frac{1}{n} \sum_{i=1}^{n} \|\varphi_i^1\|^2 \geq \lim_{n\to\infty} \frac{1}{n} \sum_{i=1}^{n} \|w_i\|^2 = \operatorname{tr} R > 0 \tag{112}$$

and

$$\varlimsup_{n\to\infty} \frac{1}{n} \sum_{i=1}^{n} \|\varphi_i^1\|^2 < \infty, \quad \text{a.s.} \tag{113}$$

From (59), (112), and (113) it follows that there are positive quantities $\beta \geq \alpha > 0$ such that

$$\alpha \cdot n \leq r_{1n} \leq \beta \cdot n, \tag{114}$$

which together with (111) yield

$$\lambda_{\max}^{1n}/\lambda_{\min}^{1n} \leq M \log^{1/4} r_{1n}, \quad \forall \quad n \geq N_0,$$

with some $M > 0$, where $\lambda_{\max}^{1n}$ and $\lambda_{\min}^{1n}$ denote, respectively, the maximum and minimum eigenvalue of

$$\sum_{i=1}^{n} \varphi_i^1 \varphi_i^{1\tau} + \frac{1}{d} I,$$

with $d = mp + lq + mr$. Then we obtain the required assertion $\Phi_1(n, 0) \xrightarrow[n\to\infty]{} 0$ by Theorem 2.2 and Remark 2.1. ■

We now specify the disturbance $\{v_n\}$ in (61).

Let $\{\epsilon_n\}$ be a sequence of independent identically distributed random vectors with $E\epsilon_i = 0$, $E\epsilon_i\epsilon_i^\tau = I$, $E\|\epsilon_i\|^{4+\delta} < \infty$ for some $\delta > 0$, and let $\{\epsilon_n\}$ and $\{w_n\}$ be mutally independent.

Define

$$v_0 = v_1 = 0, \quad v_n = \epsilon_n/\log^{1/8} n, \quad n \geq 2. \tag{115}$$

It is easy to verify that

$$\underline{\lim}_{n\to\infty} \lambda_{\min} \frac{\log^{1/4} n}{n} \sum_{i=1}^{n} Ev_i v_i^\tau \neq 0 \tag{116}$$

since

$$\frac{\log^{1/4} n}{n} \sum_{i=1}^{n} Ev_i v_i^\tau = \frac{\log^{1/4} n}{n} \sum_{i=2}^{n} \frac{1}{\log^{1/4} i} \cdot I$$

$$\geq \frac{\log^{1/4} n}{n} \int_2^{n-1} \frac{dt}{\log^{1/4} t} I \xrightarrow[n\to\infty]{} I$$

by the L'Hospital rule.

The following condition (A_6) will be used in the sequel.

(A_6) $B_1^+A(z)$ and $B_1^+B(z)$ are left coprime and $B_1^+B_q$ is of full rank.

Theorem 4.1.

For system (1) and the algorithm defined by (5) through (8), if conditions (A_1), (A_2), (A_4)-(A_6) hold and $\{v_n\}$ is defined by (115), then the adaptive tracking system has the following properties:

(i) *Stability:*

$$\overline{\lim_{n\to\infty}} \frac{1}{n} \sum_{i=1}^{n} \|u_i\|^2 < \infty, \quad \overline{\lim_{n\to\infty}} \frac{1}{n} \sum_{i=1}^{n} \|y_i\|^2 < \infty, \quad \text{a.s.,} \tag{117}$$

(ii) *Optimality:*

$$\lim_{n\to\infty} \frac{1}{n} \sum_{i=1}^{n} \left(y_i - y_i^*\right)\left(y_i - y_i^*\right)^\tau = R, \quad \text{a.s.,} \tag{118}$$

(iii) *Consistency:*

$$\lim_{n\to\infty} \theta_n = \theta, \quad \text{a.s.} \tag{119}$$

Proof. Assertions (117) and (118) are given in Lemma 4.1. Thus we only need to prove (119). For this by Lemma 4.4, it suffices to verify that

$$\underline{\lim}_{N\to\infty} \lambda_{\min}\left(\frac{\log^{1/4} N}{N} \sum_{n=1}^{N} z_n z_n^\tau\right) \neq 0, \tag{120}$$

where z_n is as defined by (91).

If (120) were not true, then there would exist a subsequence of eigenvectors $\alpha_{N_k} \in R^{mp+lq}$ for matrix

$$\frac{\log^{1/4} N_k}{N_k} \sum_{n=1}^{N_k} z_n z_n^\tau$$

with $N_k \xrightarrow[k\to\infty]{} \infty$ and

$$\|\alpha_{N_k}\| = 1 \quad \forall \quad k \geq 1 \tag{121}$$

such that

$$\alpha_{N_k}^\tau \frac{\log^{1/4} N_k}{N_k} \sum_{n=1}^{N_k} z_n z_n^\tau \alpha_{N_k} \xrightarrow[k\to\infty]{} 0. \tag{122}$$

Without loss of generality we suppose that $\alpha_{N_k} \xrightarrow[k\to\infty]{} \alpha$. We write α_{N_k} and α in the component form

$$\alpha_{N_k} = \left[\alpha_1^\tau(N_k) \cdots \alpha_{p+q}^\tau(N_k)\right]^\tau, \quad \alpha = \left[\alpha_1^\tau \cdots \alpha_{p+q}^\tau\right]^\tau,$$

with $\alpha_i(N_k)$, α_i being m-dimensional and $\alpha_{p+j}(N_k)$, α_{p+j} l-dimensional vectors, $i = 1, \ldots, p$, $j = 1, \ldots, q$. Set

$$H_{N_k}(z) = \alpha_1^\tau(N_k)z + \cdots + \alpha_p^\tau(N_k)z^p + \alpha_{p+1}^\tau(N_k)H_1(z) + \cdots + \alpha_{p+q}^\tau(N_k)H_1(z)z^{q-1}$$

$$\triangleq \sum_{i=0}^{\infty} h_i^\tau(N_k)z^i, \tag{123}$$

$$H(z) = \alpha_1^\tau z + \cdots + \alpha_p^\tau z^p + \alpha_{p+1}^\tau H_1(z) + \cdots + \alpha_{p+q}^\tau H_1(z)z^{q-1}$$

$$\triangleq \sum_{i=0}^{\infty} h_i^\tau z^i. \tag{124}$$

By condition (A_2) it is easy to see that there are constants $c_1 > 0$, $c_2 > 0$ such that $\|h_i(N_k)\| \le c_1 \exp(-c_2 i)$, $\forall\ i \ge 0$, $\forall\ k \ge 0$, whence Lemma 4.2 can be applied.

From (122), (123), and Lemma 4.2 we have

$$0 = \lim_{k\to\infty} \frac{\log^{1/4} N_k}{N_k} \sum_{n=1}^{N_k} \Big[\alpha_1^\tau(N_k)v_{n-1} + \cdots + \alpha_p^\tau(N_k)v_{n-p} + \alpha_{p+1}^\tau(N_k)H_1(z)v_n + \cdots + \alpha_{p+q}^\tau(N_k)H_1(z)v_{n-q+1}\Big]$$

$$= \lim_{k\to\infty} \frac{\log^{1/4} N_k}{N_k} \sum_{n=1}^{N_k} \Big[\Big(\alpha_1^\tau(N_k)z + \cdots + \alpha_p^\tau(N_k)z^p + \alpha_{p+1}^\tau(N_k)H_1(z) + \cdots + \alpha_{p+q}^\tau(N_k)H_1(z)z^{q-1}\Big)v_n\Big]^2$$

$$= \lim_{k\to\infty} \frac{\log^{1/4} N_k}{N_k} \sum_{n=1}^{N_k} (H_{N_k}(z)v_n)(H_{N_k}(z)v_n)^\tau$$

$$= \lim_{k\to\infty} \frac{\log^{1/4} N_k}{N_k} \sum_{n=1}^{N_k} h_i^\tau(N_k)\left(\sum_{n=1}^{N_k} R_n\right)h_i(N_k),$$

which implies via (116) that

$$\lim_{k\to\infty} \sum_{i=0}^{N_k} \|h_i(N_k)\|^2 = 0,$$

and hence

$$\sum_{i=0}^{\infty} \|h_i\|^2 = 0$$

by the dominated convergence theorem; therefore $H(z) = 0$. Setting $z = 0$ and paying attention to the fact that $H_1(0)\left(=B_1^+\right)$ is of full row rank, we see that $\alpha_{p+1}^\tau B_1^+ = 0$ and so $\alpha_{p+1} = 0$; whence it follows directly from (124) that

$$\left(\alpha_1^\tau + \alpha_2^\tau z + \cdots + \alpha_p^\tau z^{p-1}\right) = -\left(\alpha_{p+2}^\tau + \cdots + \alpha_{p+q}^\tau z^{q-2}\right)H_1(z). \tag{125}$$

In view of condition (A_6), applying Lemma 6.6-1 of [26] to (125), we know that there exists a polynomial with vector coefficients

$$f(z) = f_1 + f_2 z + \cdots + f_s z^{s-1}, \quad s \geq 1,$$

such that

$$\begin{aligned}\alpha_{p+2}^\tau + \cdots + \alpha_{p+q}^\tau z^{q-2} &= f^\tau(z)B_1^+B(z)\\ &= \left(f_1^\tau + \cdots + f_s^\tau z^{s-1}\right)\\ &\quad\times\left(B_1^+B_1 + \cdots + B_1^+B_q z^{q-1}\right).\end{aligned} \tag{126}$$

From here it is easy to conclude that $f_i = 0$ $(1 \leq i \leq s)$ since $B_1^+ B_q$ is of full rank by condition (A_6). Then $\alpha_{p+j} = 0$ by (126), and $\alpha_i = 0$ by (125) $(1 \leq j \leq q,\ 1 \leq i \leq p)$. Thus $\alpha = 0$, and $\alpha_{N_k} \xrightarrow[k\to\infty]{} 0$; this contradicts (121). Hence (120) holds. ■

Remark 4.1. It is clear that the optimal adaptive control is not unique and that v_n is not necessarily of the form given by (115). In fact, any $\{v_n\}$ satisfying (83), (116), and condition (A_4) makes Theorem 4.1 true.

Now let us remove the artificially introduced disturbance $\{v_n\}$ and see for what kind of $\left\{y_n^*\right\}$ the conclusions of Theorem 4.1 remain valid.

Setting $v_n \equiv 0$ in Lemma 4.4 and using Theorem 3.2, we obtain the following theorem.

Theorem 4.2

For system (1) and algorithm defined by (1) through (8), if conditions for $\{w_n\}$ in (A_4) and (A_1)-(A_3) are fulfilled and if Y_n^* defined by (90) satisfies

$$\varliminf_{N\to\infty} \lambda_{\min}\left(\frac{\log^{1/4} N}{N} \sum_{n=1}^{N} Y_n^* Y_n^{*\tau}\right) \neq 0, \tag{127}$$

then (117) through (119) hold.

By invoking condition (A_6), (127) can be simplified; namely, instead of (127) we shall use the following condition (a_1) or (b_1) with $H_1(z)$ removed.

(a_1) The system (1) is of single input and single output and $\left\{y_n^*\right\}$ is such that

$$\varliminf_{N\to\infty} \lambda_{\min}\left(\frac{\log^{1/4} N}{N} \sum_{n=0}^{N} Y_0^*(n) Y_0^{*\tau}(n)\right) \neq 0,$$

where

$$Y_0^*(n) = \left[y_n^*, y_{n-1}^*, \ldots, y_{n-p-q+1}^*\right]^\tau.$$

(b_1) $\left\{y_n^*\right\}$ is such that

$$\lim_{n\to\infty} \lambda_{\min}\left(\frac{1}{n}\sum_{i=1}^{n} y_i^* y_i^{*\tau}\right) \neq 0$$

and

$$\lim_{N\to\infty} \frac{1}{N}\sum_{n=1}^{N} y_{n-i}^* y_{n-j}^{*\tau} = 0$$

for $i \neq j$.

Theorem 4.3

For system (1) and the algorithm defined by (5) through (8), if conditions (A_1)-(A_3), (A_6), and (a_1) or (b_1) are fulfilled and if $\{w_n\}$ meets the requirements in (A_4), then (117) through (119) hold.

Proof. We first show that condition (a_1) implies (127). To this end, set

$$\bar{Y}_0^*(n) \triangleq B(z)Y_n^* = \left[B(z)y_n^* \cdots B(z)y_{n-p+1}^*,\right.$$
$$\left. A(z)y_{n+1}^* \cdots A(z)y_{n-q+2}^*\right]^\tau.$$

By a similar way as in the proof of (120) we can show that

$$\lim_{N\to\infty} \lambda_{\min}\left(\frac{\log^{1/4} N}{N}\sum_{n=1}^{N} \bar{Y}_0^*(n)\bar{Y}_0^{*\tau}(n)\right) \neq 0. \tag{128}$$

Now, if (127) were not true, then there would be $\beta_{N_k} \in R^{p+q}$ with $\|\beta_{N_k}\| = 1$ such that

$$\frac{\log^{1/4} N_k}{N_k}\sum_{n=1}^{N_k} \|\beta_{N_k}^\tau Y_{n-l}^*\|^2 \xrightarrow[k\to\infty]{} 0, \quad \forall \ l \geq 0. \tag{129}$$

By (128) there is $c_0 > 0$ so that

$$0 < c_0 \le \frac{\log^{1/4} N_k}{N_k} \sum_{n=1}^{N_k} \| \beta_{N_k}^{\tau} \bar{Y}_0^*(n) \|^2$$

$$= \frac{\log^{1/4} N_k}{N_k} \sum_{n=1}^{N_k} \left\| \sum_{i=1}^{q} B_i \beta_{N_k}^{\tau} Y_{n-i+1}^* \right\|^2$$

$$\le q \sum_{i=1}^{q} B_i^2 \frac{\log^{1/4} N_k}{N_k} \sum_{n=1}^{N_k} \| \beta_{N_k}^{\tau} Y_{n-i+1}^* \|^2,$$

which would go to 0 by (129). The contradiction shows that (127) holds.

To complete the proof of the theorem, we must show that condition (b_1) implies (127), but this can be done by use of a similar argument as in the proof of (120). ■

Example. Let $P(z)$ be a polynomial with rational coefficients and with order greater than 1. Then we can show that $y_n^* = \sin P(n)$ satisfies both (a_1) and (b_1).

V. CONVERGENCE RATE

In this section for the special case $r = 0$, we give the convergence rate for both the parameter estimates and the tracking error. For the proofs of these results we refer the reader to [15].

Theorem 5.1

For system (1) with $r = 0$ and the algorithm defined by (5) through (10) with any initial value θ_0, the necessary and sufficient condition for $\theta_n \to \theta$ is $\Phi(n, 0) \to 0$ and in this case

$$\| \theta_n - \theta \| = 0(\| \Phi(n, 0) \|^{\delta/1+\delta}), \quad \forall \ \delta \in [0, 1 - \epsilon/2), \quad n \to \infty.$$

We note that here the convergence rate is expressed in terms of $\|\Phi(n, 0)\|$, a quantity not easily available. We now use r_n to describe it.

Set $\tilde{\theta}_n = \theta_n - \theta$. We then have the following theorem.

Theorem 5.2

If for system (1) with $r = 0$ and algorithm given by (5) through (10) as $n \to \infty$, $r_n \to \infty$, and

$$\overline{\lim_{n\to\infty}} r_n/r_{n-1} < \infty,$$

then

(i) $\|\tilde{\theta}_n\| = o\left(r_n^{-\delta_1}\right)$, for some $\delta_1 > 0$ if $\lambda_{max}^n/\lambda_{min}^n \leq \gamma < \infty$,

and

(ii) $\|\tilde{\theta}_n\| = o\left(\{\log r_n\}^{-\delta_2}\right)$ for some $\delta_2 > 0$ if $\lambda_{max}^n/\lambda_{min}^n \leq M(\log r_n)^{1/4}$, $\forall\ n \geq N_0$, where γ, M, and N_0 all are positive quantities possibly depending on ω and λ_{max}^m and λ_{min}^n are defined in Theorem 2.2.

Theorem 5.3

For system (1) with $r = 0$ and the algorithm defined by (5) through (8), if conditions (A_2), (A_4), (A_5), and (A_6) are satisfied and $\{w_i\}$ and $\{v_i\}$ both are identically distributed sequences with $Ev_iv_i^\tau = R_1 > 0$, then there is a $\delta_1 \in (0, 1/d)$ $(d = mp + lq)$ such that

$$\|\tilde{\theta}_n\| = o(n^{-\delta_1}), \quad \text{a.s.} \quad n \to \infty,$$

and the long-run average has the expansion

$$\frac{1}{n}\sum_{i=1}^{n}\|y_i - y_i^*\|^2 = \mathrm{tr}(R + R_1) + o(n^{-\epsilon_1}), \quad \forall\ \epsilon_1 \in (0, \delta_1).$$

Theorem 5.4

Suppose that for system (1) with r = 0 and the algorithm given by (5) through (8), $\{w_n\}$ is a sequence of independent identically distributed random vectors with $Ew_n = 0$, $Ew_n w_n^\tau = R > 0$, and $E\|w_n\|^{4+\delta} < \infty$ for some $\delta > 0$ and that conditions (A_2) and (A_3) hold and

$$\underline{\lim}_{N\to\infty} \lambda_{\min}\left(\frac{1}{N} \sum_{n=1}^{N} Y_0^*(n) Y_0^{*\tau}(n)\right) \neq 0, \tag{130}$$

with Y_n^* defined by (90). Then

$$\|\tilde{\theta}_n\| = o(n^{-\delta_1}) \quad \text{a.s.} \quad \text{for some} \quad \delta_1 \in (0, 1/d)$$

and

$$\frac{1}{n} \sum_{i=1}^{n} \|y_i - y_i^*\|^2 = \text{tr } R + o(n^{-\epsilon_1}), \quad \forall \ \epsilon_1 \in (0, \delta_1).$$

Remark 5.1. If condition (A_6) holds and (130) is replaced by condition (b_1) or (a_1) with $\log^{1/4} N$ removed, then Theorem 5.4 remains valid.

REFERENCES

1. K. J. ÅSTRÖM and B. WITTENMARK, "On Self-Tuning Regulators," *Automatica 9*, 185-199 (1973).

2. G. C. GOODWIN, P. J. RAMADGE, and P. E. CAINES, "Discrete Time Stochastic Adaptive Control," *SIAM J. Control Optim. 19*, 829-853 (1981).

3. H. F. CHEN, "Self-Tuning Controller and Its Convergence under Correlated Noise," *Int. J. Control 35*, No. 6, 1051-1059 (1982).

4. P. R. KUMAR, "Adaptive Control with a Compact Parameter Set," *SIAM J. Control Optim. 20*, 9-13 (1982).

5. K. S. SIN and G. C. GOODWIN, "Stochastic Adaptive Control Using a Modified Least Squares Algorithm," *Automatica 18*, No. 3, 315-321 (1982).

6. L. LJUNG, "Consistency of the Least Squares Identification Method," *IEEE Trans. Autom. Control AC-21*, No. 5, 779-781 (1976).

7. L. LJUNG, "Analysis of Recursive Stochastic Algorithm," *IEEE Trans. Autom. Control AC-22*, No. 4, 551-575 (1977).

8. J. B. MOORE, "On Strong Consistency of Least Squares Identification Algorithm," *Automatica 14*, No. 5, 505-509 (1978).

9. H. J. KUSHNER and D. S. CLARK, "Stochastic Approximation Methods for Constrained and Unconstrained Systems, Springer-Verlag, Berlin and New York (1978).

10. V. SOLO, "The Convergence of AML," *IEEE Trans. Autom. Control AC-24*, No. 6, 958-962 (1979).

11. H. F. CHEN, "Strong Consistency of Recursive Identification under Correlated Noise," *J. Syst. Sci. Math. Sci. 1*, No. 1, 34-52 (1981).

12. H. F. CHEN, "Quasi-Least-Squares Identification and Its Strong Consistency," *Int. J. Control 34*, No. 5, 921-936 (1981).

13. H. F. CHEN, "Strong Consistency and Convergence Rate of Least Squares Identification," *Sci. Sin. (Ser. A) 25*, No. 7, 771-784 (1982).

14. H. F. CHEN and L. GUO, "Strong Consistency of Parameter Estimates for Discrete-Time Stochastic Systems," *J. Syst. Sci. Math. Sci. 5*, No. 2, 81-93 (1985).

15. H. F. CHEN and L. GUO, "Strong Consistency of Recursive Identification by No Use of Persistent Excitation Condition," *Acta Math. Appl. Sin. (Engl. Ser.) 2*, No. 2, 133-145 (1985).

16. A. BECKER, P. R. KUMAR, and C. Z. WEI, "Adaptive Control with the Stochastic Approximation Algorithm—Geometry and Convergence," *IEEE Trans. Autom. Control*, in press (1986).

17. P. E. CAINES and S. LAFORTUNE, "Adaptive Control with Recursive Identification for Stochastic Linear Systems," *IEEE Trans. Autom. Control AC-29* (1984).

18. H. F. CHEN, "Recursive System Identification and Adaptive Control by Use of the Modified Least Squares Algorithm," *SIAM J. Control Optim. 22*, No. 5 (1984).

19. H. F. CHEN and P. E. CAINES, "Strong Consistency of the Stochastic Gradient Algorithm of Adaptive Control," *Proc. Conf. Decision Control, Las Vegas, Nevada*, December 1984; also accepted by *IEEE Trans. Autom. Control*.

20. H. F. CHEN and L. GUO, "Asymptotically Optimal Adaptive Control with Consistent Parameter Estimates," *SIAM J. Control Optim. 25*, No. 3 (1987).

21. G. C. GOODWIN, D. J. HILL, and M. PALANISWAMI, "A Perspective on Convergence of Adaptive Control Algorithms," *Automatica 20*, No. 5, 519-531 (1984).

22. G. H. HARDY, J. E. LITTLEWOOD, and G. POLYA, "Inequalities," Cambridge Univ. Press, London and New York, 1934.

23. Y. S. CHOW, "Local Convergence of Martingale and the Law of Large Numbers," *Ann. Math. Stat. 36*, 552-558 (1965).

24. S. P. MEYN and P. E. CAINES, "The Zero Divisor of Multivariable Stochastic Adaptive Control Systems," *Systems and Control Letters 6*, No. 4, 235-238 (1985).

25. Y. S. CHOW and H. TEICHER, "Probability Theory," Springer-Verlag, Berlin and New York, 1978.

26. T. KAILATH, "Linear Systems," Prentice-Hall, Englewood Cliffs, New Jersey, 1980.

INDEX

R

S

T